FOOD POWER
FROM THE SEA

ALSO BY LEE FRYER AND DICK SIMMONS

Ecological Gardening for Home Foods
Whole Foods for You

BY LEE FRYER AND ANNETTE DICKINSON

A Dictionary of Food Supplements

FOOD POWER FROM THE SEA:

The Seaweed Story

LEE FRYER and DICK SIMMONS

Illustrations by Christine Becker

 MASON/CHARTER NEW YORK 1977

Library of Congress Cataloging in Publication Data

Fryer, Leland N   1908–
    Food power from the sea.

    Includes index.
    1. Marine algae as food.   2. Marine algae as
fertilizer.   3. Fishery resources.   I. Simmons, Dick,
joint author.   II. Title.
SH390.F79   639   76-47000
ISBN O-88405-383-0
ISBN O-88405-404-7 pbk.

This book is dedicated to

MALCOLM REISS

beachcomber and poet of the sea

He helped to plan it
and his spirit is found
in many of the pages.

CONTENTS

INDEX TO SPECIAL INSERTS

INDEX TO SPECIAL INSERTS

ACKNOWLEDGMENTS

Dr. Bob Wildman, Office Of Sea Grant, National Oceanic & Atmospheric Administration (NOAA), gave personal and official assistance as this book was written.

Mort Baill and Roy Rendahl of Economic Development Administration (EDA) provided encouragement and shared information from their office and projects.

Dr. T. L. Senn, Head, Horticulture Department, Clemson University, gave access to unpublished data on seaweed research as well as reports of over 15 years of studies in this field.

Dr. John Hall, Agronomist, University of Maryland, shared unpublished research data from his studies of pest control with seaweed and was always available for counsel.

Paul Wolfgang, General Manager of Sea Born, Inc., Charles City, Iowa, provided test work and pictures for use in producing this book.

Charles Walters, Jr., Editor-Publisher of *Acres, U.S.A.*, gave steady encouragement and published some of the information in earlier articles in *Acres*.

Sy Gresser, production adviser for Earth Foods Associates, helped with preparing and editing the manuscript.

Marian Fryer and Arthur Fryer helped to finance research and graphics for the book.

ACKNOWLEDGMENTS

Tandrea and Dion, our children, were cheerful as usual and helped to provide a good work environment.

Nancy Davis, our editor at Mason/Charter Publishers, gave skilled help at each stage to assure that this is a useful and beautiful book.

All of this help, and other kinds every day, were deeply appreciated by the authors.

Of the Use of Seaweeds as Manure
by
Thomas Hale, Esq.
By His Majesty's Royal License And Authority
London, 1756

The use and value of seaweeds as manure depend on the plainest principle in the World. All vegetables whatever are rich manures when in a state of decay; salt we have also seen is a very rich dressing; now seaweeds have the double advantage of their own vegetable nature, and of the seawater in which they grow.

Nay, there is something yet more than this in favor of their richness. Curious persons who have examined them according to chemistry, have found that they contain much of the same principles as animals; and it has since been discovered by the help of glasses, that they are always crowded with little insects that live upon their slimey surfaces, or in their little hollows. This is so strongly visible in many of them, that some ingenious persons both in England and elsewhere, have supposed them not to grow as plants, but that they were made by these little creatures.

In Cornwall, where the shores are sandy or stony . . . they tear off the seaweeds from the rocks and stones, and rake together such as are cast up by storms. These they lay upon the ground without any preparation, plowing them in, and they enrich it to a surprising degree.

In the first year many of the tough kinds remain almost entire in the soil, but they give a great deal of fruitfulness to it notwithstanding; the next year they generally break and rot, and they continue nearly equal in point of fertility that, and in the succeeding year.

In some parts of Cornwall, they pile them (the seaweeds) in heaps, and cover them that they may rot before they use them. This makes them take effect in a prodigious manner the first year.

From: *A Compleat Body Of Husbandry*, Chapter XVII. Provided from the Library Archives Of Canada with the assistance of Dr. Allen E. Smith, Research Scientist, Agricultural Research Station, Regina, Saskatchewan.

FOOD POWER
FROM THE SEA

1.

PORK CHOPS AND NATURAL GAS: AMERICA'S FOOD-ENERGY DILEMMA

The pork and beef you eat today
Were gas and oil just yesterday.

Stop and think. Have you read or noticed in TV ads that hair is over 90 percent protein, so it may be "fed" with protein shampoos? And that beef, pork, poultry, and fish are protein foods; so are bread, milk, and breakfast cereals?

Did you also know that nitrogen is the main element in protein, being the key building block that makes it *be* protein? And that nitrogen is also a fertilizer used to make plants grow, the most needed fertilizer of all and the kind in farm manure and composts so essential for growing good crops, lawns, golf courses, and gardens?

Did you ever connect these various thoughts together to see that hair, muscles, animals, birds, fish, grains, foods—and you, reading this book—are examples of protein, and therefore of nitrogen, the energy-giving element so essential for farming and food production in the United States and everywhere?

If you can sense and understand even 50 percent of these facts and relationships, you are on the right wavelength for reading this book, since it concerns the critical matter of energy for producing our daily foods.

1

That is why we say at the very beginning, "Pork chops and natural gas," indicating that in these latter days of U.S. agriculture, we discarded "organic" farming and went whole hog to gas and oil as primary energy sources in producing the nation's food supplies. U.S. agriculture has become a petrochemical park. Gas is used to make the nitrogen fertilizers, and oil is used to make the pesticides. Consequently, farming competes head-on with autos, home heating, transport, and urban industry for energy, not simply for fuel to run the tractors, trucks, and harvesting machines, but also for huge amounts of crop-growing energy.

When gas and oil were cheap and plentiful, this seemed to be a good food-growing technology, opposed only by environmentalists and organic food people. But, as we experienced the "energy crisis" of 1973–1975, the petrochemical system of farming became troublesome, showing extraordinary cost increases and problems due to shortages of fossil fuels.

And now, for better or worse, America is about to "go organic" again, not because the nature lovers won the battle, but because we are running out of petroleum and natural gas. It is, in a sense, a bicentennial event. For two hundred years, the country was built on a foundation of organic farms and foods, while no one questioned the farming system, the food supplies, or the basic health of the nation. Sound units of decentralized agriculture dotted the countryside from the Atlantic to the Pacific, until we had over six million independent farms in 1945, all utilizing crop and livestock wastes and the nitrogen cycle as principal sources of energy.

Then abruptly, in a short span of thirty years, we "went chemical" like gang busters, shifting agriculture's energy base from organic wastes and the nitrogen cycle to petrochemicals—to natural gas and petroleum. This shift to fossil energy has now become the basis of America's food-energy dilemma, making it mandatory that we return by all means possible to "natural" fertilizers and pesticides.

The bicentennial holiday during which we consumed huge quantities of fossil fuels will soon be over. Tomorrow we will be "drying out," returning to an updated mode of organic

2

farming and gardening in America. The Petroleum Age is dead. Long live Petroleum!—what we have left of it.

In our frantic efforts to find alternative energy supplies for growing foods, we will surely overlook nothing: chicken feathers, hair from barber shops, cannery wastes, fish guts, blood, humates, sewage, garbage, crab wastes, seaweed, old shoes . . . everything containing protein and nitrogen. All are eligible, since they can fertilize more crops and feed more people . . . as the gas wells go dry.

Getting down to cases, even *you* reading this book, are of long-range interest to us, the fertilizer and food people. You are worth about three dollars, more or less, as a high-quality organic fertilizer, and let us remember: never grows so red the rose as where some patriot died.

However, the situation is not *that* desperate . . . yet. Since 1776, Americans have used many kinds of fertilizers and food-growing technologies, beginning with the celebrated fish used by the Indians to grow maize in New England. In later chapters, we will assess the various technologies and materials that may be used in energy-saving modes of farming and gardening. We will focus on seaweed and fish as strategic resources to use in lieu of petrochemicals in land-based food production.

Meanwhile, to acquire perspective on this food-energy-cost situation, we shall describe corn and hog production on typical Midwest farms in 1937, compared with 1977, in order to see clearly that today's pork chops—and other foods—are, indeed, produced with gas and oil.

In 1937, forty years ago:

- The corn was planted in last year's hay or grain fields, using crop rotation to sustain the yield of corn.
- The corn was fertilized with farm manure and the residues of previous crops, instead of "store-bought" fertilizer.
- The hogs were grown and fattened with the low-energy-cost corn, with minimum need for off-farm energy.
- Very little pesticides were needed or used, since the

crop rotations and "natural" fertilizers helped to control insects and plant diseases.

However in 1977:

- Most of the old farms of 1937 have been consolidated into big agribusiness units.
- The corn field is a mile long, unmixed with other crops.
- The corn is fertilized with factory-made nitrogen, rather than with crop residues and animal manure.
- The nitrogen fertilizer is made out of natural gas and liquefied air (petrochemical energy).
- Insects and diseases are controlled with toxic pesticides made of petroleum.
- Often the corn does not mature properly, so some of it is dried with gas or oil at harvest time.
- The corn is then fed to hogs whose cost is firmly linked to supplies and costs of gas and oil.

This is the heart of America's food-energy-cost dilemma, and it raises two questions: "If petrochemicals are no longer available as prime energy resources for growing foods, or are too scarce and expensive, what other kinds of crop growing energy are available to take their place?" and, "Must we return to old-fashioned *organic farming* of 1937, and perhaps permanently higher food costs, or are there other alternatives?"

Before trying to answer these questions, let us measure the size of this problem.

HOW MUCH GAS IS USED FOR FERTILIZERS?

In making nitrogen fertilizers for farms and gardens, natural gas has two roles: (1) to provide hydrogen (H) in the ammonia compound (NH_3) out of which most nitrogen fertilizers are made, and (2) to provide energy to make the ammonia. In making one ton of ammonia, about 24,000 cubic feet of natural gas are needed for the hydrogen units; another 12,000

4

cubic feet for the industrial operation of making the ammonia product. Thus, a total of about 36,000 cubic feet of gas is used to make one ton of anhydrous ammonia.

Fertilizer companies then utilize the ammonia in making urea, ammo-phos, ammonium nitrate, sulphate of ammonia, ureaform, nitrogen solutions, and dozens of other kinds of farm and garden fertilizers.

To produce the U.S. corn crop, farmers use about 100 pounds of actual nitrogen fertilizer per acre. It takes about 2,000 cubic feet of natural gas to make this fertilizer for one acre of corn. For the nation's 75 million acres of corn in 1975, farmers used about 3½ million tons of actual nitrogen fertilizer, requiring over 150 *billion* cubic feet of natural gas to manufacture.

For all kinds of crops, including vegetables, tobacco, grains, cotton, fruit, turf, and home gardens, we are using about 9 million tons of elemental nitrogen in the fertilizers,[1] requiring about 360 *billion* cubic feet of natural gas to manufacture.

This is approximately the amount of natural gas needed to heat 3 million homes in the temperate climate portion of America, in such cities as New York, Philadelphia, Cleveland, Chicago, St. Louis, and Seattle.

COSTS OF NATURAL GAS

As recently as 1972, natural gas for fertilizers cost about 15 cents per 1,000 cubic feet. As will be remembered, this was prior to America's "energy crisis." Today (1976), natural gas costs about 80 cents per 1,000 cubic feet, an increase of about 500 percent.

This jump in cost was a major factor raising prices of meats, cereals, and other basic foods during 1974–1976.

Where will the cost of gas go in the future? This is a difficult question to answer, since Congress and the Federal Power Commission will strongly influence natural gas prices by their regulatory actions. However, the costs of Canadian

[1] Data from *Commercial Fertilizers* (Washington, D.C.: Statistical Reporting Service, U.S. Department of Agriculture, May, 1975).

and Mexican gas imported into the U.S. have risen to about $1.50 per 1,000 cubic feet, and we foresee a gas cost by 1980 of about $2 per 1,000 cubic feet, an increase of about 1300 percent during the 1970–1980 decade.[2]

This increase in cost will be translated into rising costs and prices of foods.

HOW MUCH PETROLEUM IS USED FOR PESTICIDES?

In a companion move toward use of petrochemicals since World War II, U.S. farmers have shifted from use of crop rotations and old-fashioned pest controls to pesticides made from petroleum, such as DDT, chlordane, toxaphene, aldrin, dieldrin, sevin, and parathion. By 1976, about 360,000,000 pounds of these oil-based chemicals were used per year for pest controls in farming and gardening.[3]

Setting aside for the moment the matter of possible damage to land, water, birds, bees, pets, and people, and contamination of foods, this much dependence on the country's petroleum supplies for pest controls in agriculture is becoming a problem. Sooner or later these questions must be answered:

- How will farmers and gardeners control insects and plant diseases when petroleum supplies decline and eventually run out?
- In the coming energy crisis, will rising costs of petroleum also raise food costs, and if so, by how much?

At present, it takes about one ton of petroleum to make one ton of technical-grade DDT, toxaphene or other oil-based pesticide. However, that is only the beginning of the petroleum requirement, since the active ingredients are diluted 80 percent to 90 percent with oil solvents to make the products sold to farmers and gardeners. Also, additional petroleum is

[2] The cost estimates in this section are based in part on records and proceedings of the Federal Power Commission, Washington, D.C., 1975.
[3] Based on *Production, Distribution, Use and Environmental Impact Potential of Selected Pesticides* (Washington, D.C.: U.S. Environmental Protection Agency, 1975). This figure is for technical-grade active ingredients.

6

used to run farm crop sprayers and dusters; and still more for dormant oil sprays on some kinds of crops.

Taking into account these various requirements for petroleum materials in an oil-based system of pest control, we believe that about 2½ million tons of such materials are being used annually for farm and garden pest control, not counting commercial, industrial, and home pesticides, such as those used to kill cockroaches and other pests in cities.

At 7 barrels of petroleum per ton, and 42 gallons per barrel, this is equivalent to about 735 million gallons of oil used per year in one form or another for pest control in American farming and gardening.

If petroleum supplies get scarce, getting sufficient amounts of these oil-based materials may become a serious problem, affecting foods and the health of the nation if alternative means of pest control are not available.

And *if* the cost of this 735 million gallons of petrochemicals doubles, or rises fivefold, this surely will raise the cost of groceries in food markets.

Again, for pest control as well as for fertilizers, farmers are in head-on competition with autos, home heating, transport, and industry for energy supplies.

EFFICIENCY IN FARMING, ENERGY-WISE, IS DECLINING

Most people see U.S. agriculture as a glistening model of efficiency, with its snorting tractors, combines, choppers, carriers, mills, dryers, elevators, airplanes, pumps, sprayers, dusters, and clod smashers. However, the basic efficiency of the industry is actually declining. Let us look at this phase of the food and energy situation.

Recently, an eminent research group at New York State College and Cornell University studied the efficiency of U.S. agriculture and produced a report entitled "Food Production and the Energy Crisis." [4] In introducing this study, the au-

[4] From *Science*, November, 1973. Copyright 1973 by the American Association for the Advancement of Science. The authors are David Pimental and J. E. Hurd of New York State College of Agriculture and Life Science; and A. C. Bellotti, M. J. Forster, I. N. Oka, O. D. Sholes, and R. J. Whitman of Cornell University.

thors said, "Fossil fuel imputs have, in fact, become so integral and indispensable to modern agriculture that the anticipated energy crisis will have significant impact upon food production in all parts of the world which have adopted the Western system."

They go on to say "As fossil fuel resources decline, the costs of obtaining fuels both from domestic and foreign sources will rapidly increase. If current use patterns continue, fuel costs are expected to double or triple in a decade and to increase nearly fivefold by the turn of the century. When energy resources become expensive, significant changes in agriculture will take place."

Selecting the national corn crop as their model for study, this research group measured the kilocalories of energy required to produce the crop, and the kilocalories in the yield of corn. With these figures they were able to measure efficiency in terms of the ratio of input to yield for selected years. The figures from 1945 and 1970 are on page 9.

Putting the message of this energy-use comparison into graphic form, we have:

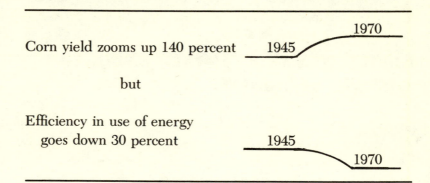

Corn yield zooms up 140 percent   1945   1970

but

Efficiency in use of energy
    goes down 30 percent   1945   1970

It is worth noting that the largest increase in energy inputs was in nitrogen fertilizer, from 58,800 kilocalories in 1945 to 940,800 in 1970—a 1700 percent increase. *If* supplies of natural gas were ample and assured for the next twenty years, and *if* the cost of gas were stable, this element in corn farming would be acceptable because the added nitrogen produces high yields. However, supplies are scarce and the cost of gas is going up, so other alternatives will have to be selected and used.

8

ENERGY INPUTS (KILOCALORIES) IN CORN PRODUCTION

INPUT	1945	1970
Labor	12,500	4,900
Machinery	180,000	420,000
Gasoline	543,400	797,000
Nitrogen	58,800	940,800
Phosphorus	10,600	47,100
Potassium	5,200	68,000
Seed	34,000	63,000
Irrigation	19,000	34,000
Insecticides	0	11,000
Herbicides	0	11,000
Drying	10,000	120,000
Electricity	32,000	310,000
Transportation	20,000	70,000
Total inputs	925,500	2,896,800
Corn yield (output)	3,427,200	8,164,800
Ratio of return to input	3.70	2.82

The corn crop only illustrates the energy supply dilemma pervading all crops and agriculture in America, including even home gardens and lawns. Good manures and composts are almost unobtainable, and 10-6-4 (percentages of nitrogen, phosphate, and potassium) in garden stores is becoming very high priced.

However, while a difficult one, this is not a dismal problem. Sensible alternatives to gas-produced nitrogen are at hand, piled high in mountains of feedlot manure, sewage sludge, and poultry wastes polluting land and water all over the country.

Recently, with the informal assistance of Dr. Robert Yeck, Staff Scientist in Waste Management, U.S. Department of Agriculture, we developed the following estimates of nitrogen supplies available in major farm wastes of America:

SOURCE	ACTUAL NITROGEN THAT MIGHT BE RECOVERED
Wastes from 11 million beef cattle	700,000 tons
Wastes from dairy cows in large dairies	700,000 tons
Wastes from poultry broilers in large colonies	250,000 tons
Wastes from laying hens	250,000 tons
Wastes from hog production	550,000 tons
Wastes from turkey-raising operations	50,000 tons
Total nitrogen available from major farm wastes	2,500,000 tons

To this substantial tonnage of nitrogen in farm wastes, we may add 500,000 tons of nitrogen recoverable from sewage sludge, making a total of 3 million tons of actual nitrogen which can be tapped intelligently—with up-to-date methods—to replace ammonia nitrogen made with natural gas.

We will explore these possibilities further in chapter 8, "Farming and Gardening in 1985 with Seaweed, Fish, and Organic Wastes."

In the meantime, let us return to seaweed and fish, since these are sleeping giants to harness and use in lieu of gas and oil as energy resources for farming and gardening.

JERSEY POTATOES

The Isle of Jersey, largest of the Channel Islands, won fame for its dairy cattle, but also for the exquisite quality of its potatoes in London markets. This, in turn, is accounted for by the use of seaweed in the Island's agriculture.

The potato was introduced in Jersey about two centuries ago, and by 1830 it was the principal cash crop. Warmer than the southern districts of England, Jersey grew the earliest of the "Earlies" available to produce handlers at Covent Garden in London; and Jersey farmers could sell all the potatoes they could grow.

To support this production, Islanders harvested seaweed from coastal shores, cutting, drying and stacking it as might be done with hay. Some was used as fuel and for cattle feed; the rest to fertilize potatoes and other crops. Interestingly, these farmers learned to grow super-early potatoes in virtually straight seaweed, mixed with a bit of sandy soil, simply covering the potato seed pieces in open furrows and letting the plants feed on this mineral-rich diet. The result was potatoes of uniquely excellent flavor, prized by Londoners, including royal families.

Soon the demand exceeded the crops so much that Jersey farmers cast abroad for new fertilizers, and found Chilean guano around 1850. It, in turn, was only another form of fish fertilizer, prepared through the mediation of fish-eating birds who deposited their guano on Chile's offshore islands. As harvested, it offered 12 percent to 15 percent nitrogen, to go along with seaweed's potash and other minerals.

Today? Well, world competition nearly wrecked Jersey's agriculture, including its fine potatoes, but the future looks bright again. Poisonous pesticides are declining; freight costs are rising. Using its rich seaweed resources, Jersey may again act as England's nearby supplier of fine potatoes.

2.

SEAWEED AND FISH IN U.S. HORTICULTURE SINCE 1876

Since Romans, Bretons, Scots, Vikings, and Spaniards used seaweed for fertilizer ever since the time of Christ, it is not surprising that Portuguese settlers near Cape Cod used this resource in growing vegetables for Boston markets. Old-timers of that area remember stories of the seaweed harvests by market gardeners after every storm, when they piled their carts high with the briny stuff to use in growing potatoes, corn, turnips, carrots, cabbages, and other kinds of produce. Their crops claimed highest prices because of the fine flavor and quality attributed to the seaweed.

Nor is it surprising that tobacco growers of Connecticut learned to use seaweed to fertilize tobacco well over 100 years ago. Seaweed is high in potash (about 3 percent) and potassium is a favored nutrient for growing fine tobacco.

The demand for seaweed must have been strong, since Luther Maddocks of Boothbay, Maine, a skilled fisherman, quit fishing in 1869 and went into production of seaweed fertilizer for sale to Connecticut tobacco farmers. He says, in his autobiography,

> That was 1869 . . . I sold out my fishing gear to the Suffolk Oil Company and decided to locate at Boothbay Harbor, where I have lived ever since.

15

My first undertaking at Boothbay Harbor was to build the Algea Fertilizer Company plant. I was making the fertilizer from dried and ground seaweed. I had a United States patent on it and a contract with the Quanipaac Company of New Haven (Connecticut) for $30 per ton for all I could dry and deliver in three years. This looked better to me than the fish business, and in the fall of '69 I built and equipped the factory which I have since used for many purposes and which is now a canning factory.[1]

Luther Maddocks then goes on to describe the difficulties of drying and grinding seaweed which, he says, "Becomes tough like leather and impossible to grind." However, he learned to cool the seaweed suddenly, then grind it into a suitable texture for use as fertilizer.

Living from 1845 to 1932, Luther Maddocks was America's first entrepreneur to harvest seaweed and process it for use in farming, for tobacco farmers of the New Haven and Hartford areas.

ATLANTIC LABORATORIES AT BOOTHBAY, MAINE

If you go to Boothbay today, four miles from Luther Maddocks's old seaweed processing plant, you will find America's new age pioneers in seaweed for fertilizers, feeds, and foods, Robert and Karen Morse, proprietors of Atlantic Laboratories, a fully modern marine products enterprise. Perhaps on dark summer nights when the moon is right, you may hear Luther Maddocks's ghost walking along the shore, smiling as he sees his old seaweed business again in operation, serving New England farmers and gardeners.

Commencing in 1971, Karen and Robert Morse harvested seaweed by hand into boats and from the beaches. Then as the demand increased, they mounted power-driven sickle bars on the boats to mow their sea-grown crops, which

[1] "Looking Backwards—Memories From the Life of Luther Maddocks." Bound typewritten manuscript in the State Library, State of Maine, Augusta.

16

Steven Cartwright

are dried and processed into seaweed meal, liquefied seaweed, and various other products. The grades for horticulture are called *Sea Crop*, those for animal feeds are *Sea Life*, and the seaweeds for human foods are called *Sea Vegetables*.

Using native seaweeds of their area and up-to-date processing facilities, Atlantic Laboratories shows good signs of growth and success. It is America's first indigenous seaweed harvester-producer for agriculture, in a new cycle of developments to serve present-day needs.

SEA-GROWN VEGETABLES FOR FISHERMEN

In the unwritten legends of West Coast fishermen it is known that a pioneer family settled in the vicinity of Queen Charlotte Island, off the coast of British Columbia around 1924. They chose a particularly rocky islet as their refuge from civilization, but it was on the route taken by fishermen and small freighters going from Puget Sound to Alaska. On their small island, these settlers developed a produce business serving boats that came their way.

Virtually without soil, they built rock embankments to hold beach sand, seaweed, clam shells, fish innards, mussels, anemones, and other rock dwellers of their shores. It took time in that cool climate for their fishy compost to become an active soil, but when it ripened it was a good one. They planted turnips, carrots, potatoes, cabbage, kohlrabi, onions, peas, parsnips, lettuce, and other vegetables not requiring too many heat units, and these did very well in the sea-fed soil. For example, the carrots grew to two and three lb. sizes, yet were tender as baby carrots. Other vegetables responded similarly. They were of superb flavor and quality.

The crews of fish-trap tenders who used these supplies found that they remained fresh and edible for a whole month while they were at sea, whereas the produce supplies from Seattle had less life and usefulness; consequently, the island produce was greatly preferred. In due time, a steady stream of towboats, seiners, tenders and gill netters patronized this ocean-side vegetable stand.

Using seaweed, fish, and sand, the family prospered,

even adding three cows and a seaweed pasture to their marine-based farm, and having all the food and income any family needs.

Now, of course, it is gone. The children went back to the city and winds and rain have returned the man-made soil to the sea.

JOSEPH V. WACHTER'S SEAWEED COMPANY— 1933 AND TODAY

Meanwhile, another American seaweed saga was in the making. It concerns Joseph V. Wachter, a Viennese immigrant and Alaskan pioneer. This man began life as a concert pianist of great promise in the era of Brahms and Dvořák; however, Wachter's health failed when he was still a young man, and he migrated to America in time to join the Alaskan gold rush.

It was a matter of kill or cure for Joseph Wachter, and he survived in Alaska as a healthy, rugged person, studying and adopting the food habits of Alaskan natives and Eskimos who prized seaweed as part of a good diet.

By 1915, fully cured of tuberculosis, Joseph Wachter was a featured pianist at the San Francisco World Fair. Then, to demonstrate his zest for life he *walked* to New York, was celebrated in that city's newspapers, and encouraged by this stroll across America, he *walked* back to San Francisco.

During 1920–1930, Joseph Wachter continued his studies of Pacific Basin seaweeds as human foods of exceptional value, even traveling to the South Pacific to observe uses of seaweed in island peoples' diets and cultures. By 1933, he had developed Organic Sea Products Cooperation at Burlingame, on the lower arm of San Francisco Bay.

First in foods, then in cosmetics and vitamins, and later in special fertilizers, Wachter's Organic Sea Products Corporation has become a successful marine enterprise, owned and operated by Joseph Wachter's sons and family.

Dr. Joseph V. Wachter, Jr. serves as President of this seaweed company. Earl Wachter, another son, manages production of the various kinds of seaweed products.

Their featured item for farm and garden use is *Sea-Spraa,* a high-quality liquid seaweed concentrate made from a blend of selected seaweed varieties.

This unique business firm has a seaweed museum and library. In it, of all things, are two bales of historic seaweed imported from Japan by Joseph V. Wachter, Sr. in the 1930s. These old seaweed packages have a "market" value of about 2,000 dollars each, because they are wrapped in pure marijuana, a customary Japanese packing and shipping material of that time.

SEAWEED AND FISH FERTILIZERS—1950–1960

World War II was finished; new kinds of fertilizers made from petrochemicals were being developed, such as ammophos, ammonium nitrate, urea, and anhydrous ammonia, to squirt into the soil and explode crops into fantastic growth. Then came the Korean War to raise prices and make fertilizers scarce.

In this era, the authors entered the fertilizer business in Seattle and Bellingham, Washington; Lee Fryer as manager of the Fertilizer Division of the Chas. H. Lilly Company, a fine old farm and garden supply firm, and Dick Simmons as proprietor of Tidewater Laboratories, Bellingham.

Liquid fertilizers were finding their way into farm and garden use. Among these, the so-called fish "emulsion" from fish meal plants showed good results but was a smelly, goopy product. We decided to make a better one.

Using fish wastes from Bellingham's fisheries, Dick Simmons developed Marina liquid fish fertilizer in a 10-6-5 grade (percentages of nitrogen, phosphate, and potash), in 1951, and we put it on the market in pretty bottles in 1952—America's first superquality, fairly odorless liquid fish fertilizer.

In this same business era, we produced Hi-Tide and Agro brands of liquid fish fertilizer, earning the top position among such products in Northwest and Alaskan horticultural use.

Seaweed meal from Norway and liquefied seaweed from England entered U.S. markets during this same period, during 1955–1960, under Algit and Maxicrop labels. We purchased and studied these new seaweed materials in the Farm and Garden Research Foundation, of which Dick Simmons was Director.

Our notes show two interesting bits of gardening performance which we present here.

GIANT CORN GROWN WITH SEAWEED

In 1959, we received four kernels of Cuzco corn seed from Peru, of a variety grown centuries ago by Inca Indians. In a plot of fertile garden soil we planted these rare seeds and blessed them with a seaweed-based fertilizer. Three of the corn seeds sprouted; the fourth was sterile and never made a plant.

Even as seedlings, these corn plants were giants. Long before the tassels came, their stalks were 3 inches in diameter, and still they grew upward.

In our mild Seattle weather, this corn grew to a height of 18 feet, with ears 9 feet above the ground. We could barely reach the bottom ear with a hoe.

Grown with seaweed fertilizer and Cuzco heredity, we believe these were world record or near world record corn plants. At any rate they illustrated an important principle in growing fine large plants: the use of complete, balanced nutrition. An ordinary NPK fertilizer (nitrogen, phosphate, and potassium) usually cannot provide these values, being deficient in many mineral nutrients; nor can seaweed alone. However, working together, these materials can produce superb yields of farm and garden crops.

POTATOES DIPPED IN LIQUID FISH

Among the technologies still unexplored in *energy-saving* food production is the impregnation of seeds with nutrients to assist new seedlings in their critical first days and weeks of life. We probed this field experimentally during 1955–1960, commencing with green peas grown by Stokeley Van Camp

21

for freezing. Since this crop is planted with over 200 pounds of seed per acre, that mass offered an opportunity to put nutrients into the seeds for early nourishment of the crop.

Using a vacuum technique, Dick Simmons removed the air from pea seed and replaced it with easily used phosphorus and nitrogen. He was able to put 6 pounds of actual phosphate into the seeds for one acre of land. The results were spectacular, giving the pea crop so much growth that it became unharvestable by regular field equipment; therefore, this technology was too early for its time. However, it may find a place in energy-saving modes of horticulture.

In 1956, using similar principles, we soaked potato seed pieces for one hour in diluted liquid fish fertilizer (Dick's Marina brand) before planting. This brief treatment, at minor cost, increased the yield in side-by-side rows by about 30 percent. This, too, may be a useful technique for increasing potato crops without much increase in fertilizer energy or cost.

OSCAR WOOD'S GIANT BEANSTALK—1975

Living in West Seattle near Alki Point, Oscar Wood has walked along Puget Sound beaches for fifty years. He is known to his neighbors and friends as an old semipro baseball player who played several seasons for the Seattle Indians; and who worked for Ma Bell's telephone company in the Seattle area for thirty-seven years.

Recently, Oscar Wood has been seeing how high he could grow a bean plant, using the Scarlet Runner variety. He does this, he says, "Mostly for fun and for seed to give away, and to see how high it will go."

In 1974, Oscar Wood's bean plant reached a height of 19 feet and its picture was shown in the *West Seattle Herald*. However, in 1975, Oscar fed his beans seaweed, and they attained a height of 24 feet.

Says Oscar Wood, "I plant two circles of seed, but of course it is one particular vine that reaches to the top. As to the seaweed, it is the green ribbonlike and ruffled variety, and sometimes the tide has left our beach covered with it. Our son-in-law raked up and loaded about six wheelbarrows

Bob Miller

full onto his trailer and put it in his compost, and he really raises a garden. One hill had forty-four large potatoes in it, besides about twenty little ones this year. We wash some of the salt water off the seaweed first. We only applied it once."

To our knowledge, during forty years of hearing farmers and gardeners tell about their big crops, Oscar Wood has the world record beanstalk. Have you seen one higher than 24 feet? If so, tell us and we will enter the achievement in the Earth Foods record book.

LEE FRYER'S BIG TOMATO PLANTS

In *Ecological Gardening for Home Foods* we described a big tomato plant 9 feet high that produced thirty delicious tomatoes.[2] However, that was before the advent of "cultured" seaweed and blends of seaweed and fish liquid fertilizers.

In 1975, we prepared the soil with seaweed meal at 1 lb. per 100 square feet, plus compost and a complete, balanced dry fertilizer. Into this soil on May 1 we planted three Better Boy tomato plants and trained them up the side of our house in Wheaton, Maryland. Each two weeks we sprayed them with a blend of Sea Born liquid seaweed and Carpole's fish.

These tomato plants attained much of their maximum potential in growth and yield of fine fruit. They grew 11 feet high and set new tomatoes steadily until mid-August, yielding about 30 lbs. per plant.

GROWING PEST-FREE CROPS WITH SEAWEED

As fertilizer formulators for twenty years, using seaweed, we have observed that plants fertilized with seaweed resist attacks by insects and plant diseases. Prior to 1970, this was thought to be interesting but less important, because the pests could be controlled with DDT, endrin, dieldrin, aldrin, parathion, chlordane, methoxychlor, sevin, and a host of other toxic pesticides.

[2] By Lee Fryer and Dick Simmons (Mason/Charter Publishers, Inc., 1975).

25

However, the ugly truth is being disclosed: many of these pest-control poisons are bad for humans, and they will kill *us* if we continue to use them in growing food crops. Therefore, there is a keen rising interest in alternative nontoxic ways to control beetles, aphids, thrips, bugs, worms, nematodes, flies, viruses, molds, mildew, and other pests that destroy food crops while they are being grown.

Seaweed and fish have major roles in such nontoxic modes of farming and gardening, and we shall discuss these benign pest controls in the pages that follow. Meantime, we will describe two examples of pest control with seaweed, as we have seen them in action.

CUCUMBERS AND SQUASH IN HALIFAX, VIRGINIA—1971

In 1971, the Rural Advancement Fund (Charlotte, N.C.) provided a small farmers' vegetable-growing cooperative near Halifax, Virginia, with financial and technical assistance. As part of this program, they wished to grow cucumbers and summer squash for the expanding higher-priced organic food market, and they called on us for nontoxic pest controls for these special crops.

We achieved this on 10 acres of "organic" production, in the midst of their 400 acres of commercial cucumbers and summer squash for other markets, by using seaweed fertilizers.

We provided 200 pounds per acre of seaweed meal, mixed into their dry fertilizer, plus two spray applications of liquefied seaweed during growth of the crops. A few bugs came and lived in these fields, but they were docile and produced no population explosions. The full-nourished plants simply outgrew the bugs and suffered virtually no damage. The seaweed control was more effective than conventional pesticides used in the adjoining fields, and a superior yield paid for the seaweed as a combination fertilizer and pest control.

27

GLEN GRABER'S PEST-FREE FARM IN OHIO

In 1972, the authors were employed by the Office of Sea Grant, U.S. Department of Commerce, to review American experience using seaweed for control of farm and garden pests; and then to prepare proposals for research in this field. Our work soon took us to Glen Graber's big farm in Ohio, and the following is what we found.

Knowing good land when he saw it, Jacob Graber, Glen's father, settled on a section of rich organic soil near Hartville in 1911, and progressively cleared 500 acres for vegetable production. During forty years he built this business into one of America's fine produce companies, known in Midwest and Eastern markets as a supplier of superior vegetables and fruits. Glen Graber took over the operations in 1953.

Like other growers, Graber commenced using the new petrochemical pesticides as they were developed and released for farming use, with the blessing of the U.S. Department of Agriculture. However, as Glen Graber told us:

> One of my small pleasures was walking through the fields, picking a sprig of celery, a lettuce leaf, a radish, or some other morsel and savoring its familiar taste.
>
> But around 1960, I began to be afraid to do this, because I knew about the poisonous materials sprayed or dusted on those crops. I began to wonder: Will this give me cancer, hurt my liver, or shake up my nerves? I quit tasting my vegetable crops, and I was old-fashioned enough to dislike selling food crops for other people to eat that I did not wish to eat myself. I went in search of better and safer pest controls.
>
> Around this time, Dr. T. L. Senn at Clemson University commenced his research on seaweed, and I went to see him. As a result, I was able gradually to eliminate all poisonous pesticides from my farming operations, using seaweed as the main pest control material.

28

When we visited Glen Graber's place in 1972, we walked through his potato, tomato, celery, radish, cabbage, and turnip fields. Some bugs were there, all right, as always happens—and will happen—in nature. But these insects were fairly scarce and hardly damaging any plants or crops. The fat, happy potato bugs just walked around, flea beetles scarcely hopped, and the aphids were not having very many children.

Through a sound system of farming, using seaweed to assure total nutrition of his crops, Glen Graber had achieved pest control without use of toxic pesticides.

Since 1972, he has had to convert his operations from vegetable growing mainly to corn and grain crops, just as other Ohio vegetable growers have done, because of the destruction of markets for independent growers by the big agribusiness units of Texas, California, Arizona, and Florida. Nevertheless, Glen Graber is America's best example of a successful large farmer who learned to use seaweed for effective pest control in vegetable crops.

Research leaders of the U.S. Department of Agriculture and Environmental Protection Agency have been strangely uninterested in such demonstrations of nonpoisonous pest control as at Graber's and in the small farmers' operations with cucumbers and summer squash. In one interview, a USDA chief in the Agricultural Research Administration said, "If I thought this was true, I'd go out to Ohio to see it."

We were reminded of an earlier Ohio event when Orville and Wilbur Wright were inventing the airplane. The editor of their Dayton newspaper would courteously ask Orville, "Well, did you fly her today?" And Orville would say, "Yep, nearly half a mile." But the newspaper never published a story about this, because the editor did not believe it was happening.

In the case of seaweed, there is substantial evidence that seaweed materials will, in fact, assist plants and crops in resisting damage by insects and plant diseases. The suggested procedure is to fertilize with 200 to 400 pounds of seaweed meal per acre (½ lb. to 1 lb. per 100 square feet), then spray the plants during growth with liquefied seaweed. Evidently, this provides a full assortment of minerals needed by the plants to mobilize their enzyme and hormone systems for self-

29

defense. This, or their rampant health due to complete nutrition, repels the insects and disease pathogens.

We are taught in historical geology that all mountains eventually wash, grain by grain, into the sea. The remains of plants, insects, and animals—all living things—go there too, adding their contributions to the mineral broth of the ocean as nutriments for growing seaweed and fish.

So, a fish is a particle of a mountain or of a baboon's whisker. A frond of seaweed has a bit of Pharaoh's farm, or an atom of camel manure. The fish and seaweed contain assortments of all minerals known to be active in life systems of our planet, useful in forming the tissues of plants and animals.

In Nature's gardens and primitive agriculture, the plants are not malnourished, because they evolved from the kinds of diets locally available to them; thus, we find salt grass in the alkaline desert and acid-loving rhododendrons in the composts of forest litter. Either one would die of malnutrition in the other's place.

However, in "advanced" civilization, we put all kinds of plants in strange places, and out of contact with their former neighbors; growing grass without a legume nearby, and 5 miles of corn alongside 5 miles of corn, alongside 5 more miles of corn. Hardly Nature's way of growing corn.

"Modern" farming oversimplifies crop nutrition to get short-term gains. For example, in growing three vegetable crops in high-speed succession on a single acre per season in Florida or Texas, it is hardly possible for the plants to gather enough minerals to form normal tissues. We suspect that this kind of agriculture disarms the plants, removing their ancient defenses against insects and plant diseases.

If so, seaweed and fish may restore the full mineral supplies, thereby helping the plants to attain normal growth, as well as providing better nutritional values for the people who consume them.

In the chapters that follow we shall describe these marine resources for land-based food production, and see how they may be used in energy-saving modes of farming and gardening, including their use to protect plants from damage by insects and diseases. We hope these guides may be useful on your own farm or in your garden.

WHAT HAPPENS WHEN SEA LEAVES FALL?

Since childhood, everyone knows that leaves and limbs of land plants slowly rot when they fall, and become food for new plants. In other cases, eaten by animals, birds, or people, land vegetation has a similar destiny: it eventually feeds new generations of life.

But how about sea plants? What is their cycle? What happens to them?

For many larger seaweeds such as laminarians and macrocystis, the fronds (leaves) do not fall. They erode at their upper edges about as fast as they grow, releasing tissues into the surrounding seawater. Other seaweeds tear loose and add to a vegetative mass suspended in the ocean.

In any case, about 20 percent of this matter (dry basis) is cellulose requiring bacterial action to convert it into food for other marine organisms. Luckily, a bit of protein is present, too, to attract and feed an initial population of marine bacteria. Busily at work, these first-feeders draw more nitrogen from the surrounding seawater, infusing it and other nutrients into the seaweed fragments.

Thus, the yeast for Neptune's grand life systems is provided. Schools of plankton eat the bacteria-enriched seaweed debris, returning their wastes to the water to be recolonized by more bacteria—and on and on in cycles of decomposition, food making and new life. Fishes, shrimp, krill, barnacles, sea anenomes, bigger plankton and even whales eat the little plankton; then all the fishy plants and creatures feed each other, casting their manure, fertilizing everything: nourishing, killing, dying, rising again in new incarnations . . . forever.

Meantime, the crafty sea urchin says, "To heck with the plankton." It pastures directly on seaweed, thus becoming seaweed's main enemy. A herd of sea urchins can mow off a field of young seaweed in only a few days.

3.

AN ENERGY-SAVING GARDEN

In America's Petroleum Age (1945–1980), hell-bent-for-bigness, government and business policies have rubbed out 5 million independent farmers, leaving less than ½ million who might improve their ways of fertilizing and growing food crops. Within this residue of farmers, only a few are strong enough financially and market-wise to serve as pioneers in energy-saving modes of agriculture.

A farmer would have to share in the Department of Agriculture's torrent of crop payments and subsidies in order to finance alternative modes of farming—using less gas and oil—and such payments are not yet available. Consequently, U.S. agriculture is a static area of development, or worse. The safety and quality of foods are still declining, and agriculture is running out of gas.

Therefore, let us agree for the moment that "small is beautiful" and use home gardens as models to describe seaweed and fish as energy-savers in producing food crops. Among America's 36 million gardeners, many are eager to learn new ways to success in raising safe, nutritious, fine-flavored vegetables and fruits. Within this group are many wives of administrators and research leaders in the Department of Agriculture and Environmental Protection Agency. Some of these will surely say to their eminent husbands,

33

"Hey, look at these gorgeous cucumbers . . . and no bugs. Why don't you do it that way in USDA and EPA?"

Or, a young seventh grade gardener will say to his food company father, "Hey dad, better come down and look at our garden. The store ran out of regular fertilizer but we *sprayed it on,* and you should see those tomatoes and beans!"

Via these roundabout routes, we may communicate in time with some farmers who are short of low-cost fertilizers and smart enough to raise crops for consumers who want safer, better-tasting foods.

Where shall we start?

Let's start in Denver, Colorado, in 1950, when the Korean War had cut off fertilizer supplies for many farmers. On reading about this, my impressionable neighbor, Bud Gregory, planted vegetable seeds in his flower bed as a measure to avert impending starvation, and his prize crop was watermelons. Imitating Nature, Bud fertilized the soil with small fish he caught in the Platte River, then "tromped" the seeds into the ground as might be done, he said, by the hooves of an antelope or the paws of a big brown bear.

Sure enough, the watermelons sprouted strongly; and short of space, Bud trained them up the wall of his house next to the kitchen door. He further fertilized them with milk and table scraps, and how they grew! (Milk contains lime, phosphorus, and nitrogen).

In July, the little watermelons came. Bud thinned them to a practical number and built a shelf upon which each could grow. One above the other up the side of his house, they were the marvel of the neighborhood. That was Bud Gregory's antistarvation energy-saving garden of 1950, thirty years early but demonstrating the ingenuity some people possess in coping with their problems. If seaweed had grown in the Platte River, we are sure Bud would have used it.

THE MODEL GARDEN

This model garden is approximately 14x30 feet in size, making 420 square feet, which is about 1/100 of an acre of land. We have selected this size for several reasons:

- Eight or ten million Americans have gardens in this size bracket, between 300 and 800 square feet.
- In such a garden, you can produce over $200 worth of vegetables and other home foods, a practical objective for many families.
- At 1/100 of an acre of land, it can serve as a "micro-farm," for farm-minded readers who think in terms of "fertilizing at 1,000 pounds per acre," or "using 100 pounds of actual nitrogen per acre." We can readily translate the garden guides into those terms.

This garden is located in a southeast exposure, where it receives at least 12 hours of sunshine per day in the spring–summer growing season. Being placed against the wall of the house and occupying a former flower bed and lawn area, the garden gets some added heat units from reflection of sunlight. Tomatoes, pole beans, and other tall plants can be trained against the house.

The model garden is in a temperate climate zone with annual rainfall of 30 to 35 inches. Percolated for a million years with rainwater, the soil has lost much of its original calcium, magnesium, phosphorus, and other minerals. Consequently, like most gardens, this one has medium to low native fertility. It needs soil-building and supplemental fertilizers in order to produce good yields of vegetables and fruits.

BUILDING AN ENERGY BASE IN THIS GARDEN

The alternative to store-bought nitrogen from natural gas is a soil capable of capturing its own nitrogen from the air and of releasing nitrogen from its own organic materials.

This means a *live* soil. What is a *live* soil? It is one with a high population of useful soil bacteria. Let us fix these linkages firmly in mind:

soil fertility equals plenty of soil bacteria;
plenty of soil bacteria equals home fertility supplies;
home fertility supplies equals energy saving; and
energy saving equals successful gardening.

Stop and think for a moment. What is a dead soil? It is

one such as might be baked in an oven, containing no living organisms . . . no bacteria, molds, worms, algae, bugs, eggs, or other living things. Dead.

Plants cannot grow in such a soil, and in Nature the soil bacteria have leading roles in feeding the plants:

- capturing nitrogen from the air and passing it along to the plants.
- decomposing organic matter and recycling its nitrogen and minerals to new generations of plants.
- breaking down soil particles and releasing their minerals to the plants

Without the services of the teensy soil bacteria, none of these biological events can occur. The soil is dead. Plants cannot grow. Therefore, as energy-saving gardeners, let us focus our thoughts on beneficial soil bacteria and on how to feed and encourage large colonies of them in our soils.

First of all, this is an unseen corner of Nature's wonderland: soil bacteria. According to Sir E. John Russell, one of the world's great horticulturalists, one gram of fertile soil may contain about 3 billion beneficial bacteria, such as the *azotobacter* or *clostridia* types.[1] This means that an acre of fertile soil may be the happy home of 3 *tons* of bacteria, live weight; and their bodies, themselves, would contain about 130 pounds of actual nitrogen, ready for use by growing plants.

WHAT IS A FERTILE SOIL?

In *Soil Conditions and Plant Growth*, Sir John Russell said,

Soils differ from a heap of inert rock particles in many ways, but one of the more important is that they have a population of micro-organisms living in them which derives its energy by oxidixing organic residues left behind by the plants growing on the soil or by the animals feeding on these plants. In the final analysis the plants growing on the soil sub-

[1] E. John Russell, *Soil Conditions and Plant Growth* (London: Longmans, Green & Co., 1950).

36

sist on the products of microbial activity, for the micro-organisms are continually oxidising the dead plant remains and leaving behind, in a form available to the plant, the nitrogenous and mineral compounds needed by the plants for their growth. On this concept a fertile soil is one which contains either an adequate supply of plant food in an available form or a microbial population which is releasing nutrients fast enough to maintain rapid plant growth; whilst an infertile soil is one in which this does not happen, as, for example, if the micro-organisms are removing and locking up available plant nutrients from the soil.[2]

Our model garden, fertile and teeming with life, will have a bacterial population of about 25,000,000,000,000,000, with a live weight of about 60 pounds. Fed and encouraged by the gardener, these busy workers will capture about 1 lb. of actual nitrogen per year from the air, and make it available to the growing plants. This is about the amount in a 20-lb. sack of 5-10-10 garden fertilizer (5 percent of 20 lbs. = 1 lb.); and the rate used by a Midwest corn farmer raising a 100 bushel per acre yield of corn.

Because they *eat* organic matter and are biological partners of the plants, the soil bacteria live in huge numbers on plant roots and in the organic portion of the soil. Sir John Russell says (above citation) that bacteria comprise about 1 percent of the weight of organic matter, they congregate on roots, acting as food gatherers and nursemaids in feeding the plants. In some cases, the plant roots are fully covered with bacterial and fungal populations, which gather foods and nourish the plants.

LOVE THOSE BACTERIA!

Now, building an energy base in the model garden is a matter of providing those foods and conditions needed by the

[2] See previous citation

soil bacteria. Love those bacteria, and they will help to fertilize the vegetable plants.

Mainly, the soil bacteria need these conditions and foods:

- **Acidity level.** They prefer a soil environment of pH 6.0
  to 7.0.
- **Lime.** The supply should be adequate, and adding lime
  will enable adjusting the pH (acidity) level.
- **Organic matter.** Adding manure, compost, or other
  organic materials will provide food and natural homes
  for bacteria, increasing their colonies and beneficial activities.
- **Air.** Bacteria need air. Adding lime, organic matter,
  and seaweed helps to regulate the air supply in the soil.
- **Minerals.** The bacteria need all kinds of minerals
  required by plants and living organisms. Seaweed and
  organic matter will supplement mineral supplies from
  the soil.

Putting this knowledge into action in autumn or early
spring, we apply the following materials to the model garden
area:

- **Dolomite-type lime,** 30 lbs. on the 420 square foot area
  (about 7 lbs. per 100 square feet). On average loam
  soil, this will raise the pH (acidity) level to about 6.8.
  We use *dolomite* type lime because it contains magnesium as well as calcium, a desirable feature.
- **Manure or compost,** 100 lbs. semidry weight. Use
  whichever is available, or both. If the supply is wet,
  such as from a nearby dairy or feedlot, use 300 lbs.
- **Seaweed meal,** 4 lbs., to provide a full assortment of
  minerals and to improve the condition of the soil. It
  will soften clay soils and help them to crumble, or if the
  soil is sandy, the seaweed provides colloids to alleviate
  that problem. Wet seaweed from an ocean beach may
  be used. If so, add it to the compost pile and compost
  it.

Spread these materials in the garden area and spade or
rototill into the top 8 inches of the soil. This is the energy
base for the model garden.

Each year, this fertility base should be replenished and
improved in autumn, the grower adding sensible amounts of

the same kinds of materials, depending on the particular soil and its performance.

On a sunny day in October or November, we clear away all of the old vines, stalks, tops, and bushes and put them in the compost pile. Harvest the last of the carrots, turnips, parsnips, and other late vegetables. This leaves the garden area clear for spading again; and we add the sensible amounts of lime, compost or manure, seaweed, and anything else locally available to promote fertility in the garden.

Phosphate may be added, with a choice of four kinds: ground rock, colloidal, bone meal, or superphosphate. All are good and a suitable amount is 10 lbs. for rock or colloidal; or 5 lbs. for bone meal or superphosphate. These are suitable applications on 420 square feet of garden space. These autumn applications, plus spring fertilizing, will sustain phosphate supplies for both soil bacteria and the garden plants.

As these words are written in December, 1975, our garden is fertilized and spaded in this preseason fashion, getting ready for spring, 1976. Even in the cold of winter the juices of seaweed, compost, lime, manure, and phosphate are percolating into grains of soil. Our 25 trillion trillion bacteria are sleepily savoring the rich mess, slowly changing fibers into black humus. By springtime, this soil will be as alive as grandma's bread dough, or as the mash in a Kentucky moonshiner's still.

PLANTING THE GARDEN VEGETABLES

In our climate zone, late February and March have warm days. The soil gets mellow and on such a day we spade or till the garden again before planting the early crops, such as peas, lettuce, and early potatoes. Meantime, we have prepared a deluxe garden fertilizer using the following ingredients:

A DELUXE GARDEN FERTILIZER

20 lbs.   Any good organic-based fertilizer available in garden stores of the area, such as a 5-8-5 rose fertilizer, or a 5-10-10 grade

SAP TESTING

The sap of a plant is like vital materials on a conveyor belt moving to sites of use in a factory; and, when a vital material is missing or in low supply, production slows down, and ceases.

Luckily, we know how to check the sap of plants quickly and cheaply to see what nutrients are there amply, in excess, low, or missing entirely. It can be done with an expanded Purdue University sap test kit such as our group used during 1953–1963 to run over 3,000 such tests out in fields of many farms and nurseries of the Northwest and Alaska.

Using such equipment, a whole farm, nursery, or garden can be covered in less than an hour. A fertilizer field worker visits the place and helps to gather samples of tissues of growing plants. He then makes little salads, extracts the sap, and checks it for nitrogen, phosphorus, potassium, magnesium, iron, and other essential nutrients. The farmer, nurseryman or gardener can then make adjustments in the fertilizing program to correct last year's mistakes, improve yields and quality, and save money.

We simply utilize the growing plant as the ultimate soil-testing mechanism, and say: "For what does it profit a farmer if he puts 200 pounds of phosphate in his soil, if it cannot enter his plants?" And, what could keep the phosphate out? Well, chlorine from too much muriate of potash (KC1) for one thing; and, ammoniation of the phosphate in making pelleted fertilizers, for another.

Now another objective can be added for using sap tests: energy saving. For example, if urea nitrogen is not getting into the plants because the soil is cold, the sap test will show this, and the farmer can switch to calcium nitrate for early season feeding, or provide part of the nitrogen via foliar sprays. This way, he saves money and nitrogen, too.

5 lbs.	Ureaform or organiform (See Appendix for description)
5 lbs.	Bone meal, or superphosphate if bone meal is unavailable
4 lbs.	Sulphate of potash magnesia (Sol-po-mag) if we can get it (See Appendix)
10 lbs.	Seaweed meal
10 lbs.	Dry manure or compost
1 lb.	Fritted trace elements (FTE), if we can get it (See Appendix)
50 lbs.	Total. Mix well *

This deluxe fertilizer can be used throughout the season in the garden and also on ornamental plants of all kinds. It is a superb rose, rhododendron, shrub, and tree fertilizer.

The seaweed-based fertilizer serves as our vegetable planting material to be put under the rows, hills, and transplants of various kinds of vegetables to assure excellent growth, yields, and quality. These are suitable methods and rates of use:

For peas, beans, carrots, greens, lettuce, onions, and other crops sown in rows, apply about 6 ounces of fertilizer (3 small handfuls) per 12 feet of row. Make the furrows an inch deeper than usual and sprinkle the fertilizer in the furrows. Mix and cover gently with soil and plant seeds as usual above the fertilizer.

For cucumbers, zucchini, melons, squash, corn, pole beans and other hill-type vegetables, put a bit of fertilizer under each hill, mix and cover with soil, and plant as usual.

For tomatoes, peppers, cabbage, broccoli, eggplants and other transplants, dig holes a big deeper than usual, add a bit of fertilizer, mix and cover with soil and proceed with planting.

This method assures complete and balanced nutrition of the vegetable crops, including full assortments of minerals in

* If you cannot find all of these ingredients, substitute or go ahead without them. You will still have a superior fertilizer and a good garden.

41

the seaweed, manure, and compost materials. Equally important, these foods nourish and support soil bacteria, which, in turn, capture nitrogen from the air and liberate nutrients from soil particles for use by the plants. These processes serve as energy-savers, reducing the need for purchased NPK fertilizers in producing high yields of food crops.

USING LIQUID SEAWEED AND FISH FERTILIZERS

Planted in fertile, fertilized soil, the model garden will produce good yields of superquality vegetables if we irrigate it and let Nature take its course. However, three additional objectives may be attained by use of liquid seaweed and fish, sprayed on the foliage of the plants:

1. *Pest control*, via positive full mineral fertilization.
2. *Maximum nutrition* and flavor in the vegetables for people who will eat them.
3. *Maximum yield,* getting high returns per unit of energy spent in production.

In later chapters we will describe insect and disease control with use of seaweed materials. Evidently, the provision of full assortments of minerals in the seaweed enables the plants to mobilize their enzyme and hormone systems for defense against pests. The evidence shows that seaweed helps to control aphids, mites, flea beetles, molds, and many other kinds of plant pests.

As known at present, the guides are as follows:

• Apply 200 to 400 pounds per acre of seaweed meal as a soil improver. We have done this in the garden program described above.
• Spray the plants with liquefied seaweed two or three times during the growing season, each time enough to wet the foliage.

In our own experience this kind of pest control has been adequate for all kinds of garden insects and pests except cabbage loopers, cutworms, slugs, white flies, and some colonies of aphids. For these stubborn local pests, we use thuricide, stale beer, plant stem guards, pyrethrin or an occasional bit of malathion—the nontoxic and/or quickly decomposing kinds of auxiliary materials. The need is minimal, since the seaweed reduces pest populations by at least 75 percent, and the plants usually outgrow the insects, thereby avoiding serious damage.

In 1975 and 1976 we used the combination of liquid seaweed and fish, blended in a 6-3-3 grade (6 percent *Ni*trogen, 3 percent *P*hosphate and 3 percent *K*-potash). This superb product fertilizes the plants, mineralizes the crop, and provides a measure of pest control, helping to attain all three of the objectives listed above.

However, the objective of energy saving is also served. Liquid nutrients supplied to foliage of plants are usually more efficient than soil-applied nutrients, because soil chemicals immobilize and lock up such plant foods as phosphorus and iron, and soil-applied nitrogen often leaches away before it can be used by the plants. Generally speaking, plant nutrients sprayed on the foliage have an efficiency at least double that of conventional dry fertilizers applied to soil.

As nitrogen costs rise along with gas and oil, the efficiency factor becomes significant. By using liquid seaweed and fish together with compost and manure, gardeners will be able to eliminate the need for other purchased fertilizers in growing good yields of garden vegetables. It is a complete, balanced supplemental fertilizer.

Here is a summary of guides for developing and fertilizing an energy-saving garden:

GUIDES FOR AN ENERGY-SAVING GARDEN

1. *Build an Energy Base*
    Add organic matter
    Love and feed the soil bacteria
    Add lime to adjust the pH in acid soils
    Aerate the soil with timely cultivation
    Use seaweed to provide plenty of minerals

2. *Use Lesser Amounts of a Good Soil Fertilizer*
    Broadcast part of this fertilizer and work it into the soil
    Add fertilizer under the rows and hills

3. *Use Liquid Seaweed and Fish as Energy-Saving Sup-plements*

Spray-feed the plants several times as they grow and produce their crops.

Use seaweed sprays to reduce pest damage and to reduce need for toxic pesticides.

AMERICA'S 36 MILLION GARDENERS GROW $7 BILLION IN FOODS

Our neighbor, Roy Priest, is a typical new-style city gardener. He says, "Why should I pay 60 cents a head for lettuce and 50 cents a pound for tomatoes when I can grow one-fourth of all the lettuce and tomatoes our family needs. And radishes, corn, beans, and peas, too."

Then he says, "This way I get low-cost recreation and exercise at home and the vegetables taste better. At least, I know they won't poison me."

Multiply Roy Priest by 36 million and you have America's gardening community. It is idealistic, up to a point, but very practical minded. In fact, gardening is like mowing lawns. The idea is good, but many people hate doing it. They would rather hire the neighbor's children for $3 or $5 per mowing.

The National Garden Surveys of 1973 and 1975, conducted by Gallup Poll for *Gardens for All, Inc.*, a nonprofit organization based in Shelburne, Vermont, show that we have 36 million gardeners, and the number is rising. Half of all American households had a vegetable garden of some size or kind in 1976, in a plot, pot, patio, window box, or green-house.

Another 10 or 15 million people would grow gardens if they could get access to land, and if they could save at least $200 in food costs by being gardeners.

However, the main problem remains: how to be success-ful. Thousands of gardeners are like our friend Roy Priest. They want the good yield of juicy, full-flavor vegetables, but they fail to harness their soil bacteria and they don't use enough fertilizer. They provide about one-fourth as much plant food as a successful commercial vegetable grower uses in his operations.

One way is *organic* gardening, but that way was best in the old mixed farm and city communities, where there was a dairy or a poultry farm nearby, and you could get plenty of farm manure on a sunny spring afternoon.

No more. The dairy man or egg supplier may be 1,000 miles away. So we have a problem. Thirty-six million Americans want to be successful gardeners, and they need near-at-hand materials and guides. Seaweed and fish materials may help many of these new gardeners to be successful.

The National Garden Surveys conducted by Gallup and *Gardens for All* show that a large and rising number of Americans are into gardening in earnest, with the aim of producing a substantial amount of their food needs. About 18 million cultivate 500 square feet or more of garden area, and 6 million of these aim to grow and store at least $1,000 worth of home foods per year.

These reports mean that home gardening is a major expanding enterprise in America. Six million gardeners at $1,000 each are $6 billion worth of home-produced foods. Add 12 million families with sizable gardens and 18 million with little gardens and this is a $7 billion enterprise in home-produced foods.

It is a big anti-inflation force. The home-produced foods help to keep prices under control. At 36 million gardeners on the way to 40 million, this is America's largest mutual interest group, wanting to be successful, needing to be served.

THE ENERGY-SAVING FACTOR

If our neighbor Roy Priest is to be a successful gardener, he needs to obtain and use, one way or another, at least 1 lb. of *actual* nitrogen a year in his 400 to 500 square foot garden. This equals the guide used in successful farming. To grow 100 bushel yields of corn, use 100 lbs. of actual nitrogen per acre, and Roy's garden is 1/100 of an acre.

If 18 million gardeners buy this much nitrogen fertilizer, it will require about 230 million cubic feet of natural gas to manufacture. The supplies are getting scarce and the cost is rising.

Based on the model energy-saving garden, we suggest cutting the need for purchased nitrogen by 50 percent in the coming five years. Then, as petrochemicals become too scarce and precious to use for making fertilizers and pesticides, shift completely to organic wastes, seaweed, and fish. As you will discover, this presents no insurmountable problems. The supplies are good and the technologies are sound.

Going this way is America's route to cheaper, safer, more healthful foods.

FERTILIZING LAND PLANTS WITH SEA SOLIDS
U.S. PATENT NO. 3,071,457—DR. M. R. MURRAY (1963)

Sea solids are those portions of seawater remaining after the water is removed; namely, the minerals. America's leading investigator of these, as to their values for horticulture, is Dr. R. M. Murray of Fort Myers, Florida, holder of patent No. 3,071,457, describing use of sea solids to fertilize land plants.

This document reports results when dried sea salts are used in varying amounts from 550 pounds to 2,200 pounds per acre in growing corn, oats, soybeans, and vegetable crops.

The observations include an increase of vitamin C in tomatoes from 100 milligrams per 100 grams of dry tomato solids in the untreated portion of a field to 245 milligrams in the plot fertilized with 1,100 lbs. per acre of the sea minerals.

Day-old chickens were divided into two groups; one fed on corn and oats from treated plots, and the other on corn and oats from untreated areas. Both were given identical rations of commercial protein, and vitamin and mineral supplements. The birds on the experimental grain supplies matured earlier than those from untreated corn and oat crops and the hens began laying eggs a month earlier. Also, the eggs were larger.

It was noted that the sea mineral fertilization influenced the incidence of disease in some of the crops. For example, corn smut was 3.8 times as prevalent in the untreated area as in the area fertilized with sea solids, and "center rot" in turnips fell from 30 percent in the untreated plot to 0 percent where the sea fertilizer was used.

The high saline content of the sea solids, particularly of common salt (NaCl), limits the amount per acre that can safely be utilized. Dr. Murray foresees a wide use of sea solids in hydroponic and hydroculture production of crops, as well as in conventional farming and gardening.

4.

SEAWEEDS FOR USE IN FARMING AND GARDENING

Many years ago, as young people, we explored the shores and beaches of Oregon, near Pacific City where the Nestucca River leaves the Coast Range mountains and flows into the sea. To the north, Cape Kiwanda, with a head like a chess rook, made a small harbor for deep-sea fishermen. Sea roses, sea urchins, starfish, and rock oysters lived at the toe of the Cape and wild huckleberries covered its slopes. It was a wild, unspoiled place.

When the roar of the surf moved south to the river's mouth, old fishermen said, "There's going to be a storm," and they were always right. Breakers climbed the sand dunes with foamy power. Afterward, we walked on the beach and gathered seaweed. Cows ate it and some people used these storm-cast weeds to fertilize their sandy gardens so they could grow tall crops of Telephone peas in springtime.

One of these early gardeners was Mr. Sukkula, a mystery-man bachelor who loved company and always invited us to eat with him. However, we accepted only once because Mr. Sukkula served us clams boiled in huckleberries.

In that era, a family of four could live quite well on the Oregon coast for five hundred dollars a year: fish, venison, oysters, clams, milk, cheese, butter, eggs, wild ducks, good water, clean air, ocean storms and beachcombing. What a paradise! That was social security.

49

WHAT IS SEAWEED? HOW DOES IT GROW?

Land people learn about land plants early in life by living and walking around among them. By the age of ten, most children know that trees grow tall, bushes are medium and grass is low and everywhere. They can tell the difference between roses and strawberries: between maple and pine trees. Adults know that sagebrush grows in dry desert places and cattails in wet places, while mountain tops 15,000 feet high grow no trees at all. They are above the timberline.

But about seaweed? What do land-livers know about seaweed? Even the name is wrong. Weeds are often defined as plants growing in the wrong places; but seaweed grows in the right places, and is useful to people. These are ocean plants, not weeds, although in deference to two thousand years of custom we shall still call them seaweeds.

Literally, seaweed is any plant that grows in ocean water, deriving nutrition directly from the water into its tissues, rather than from soils through roots. Sea plants have no roots or circulatory systems. They feed from the seawater that surrounds them.

The seaweeds attached to the bottom have holdfasts, rootlike organs that cling to rocks, but they are not food-gatherers. They are simply mechanical devices to hold the plants in place against tides and storms. However, some kinds of seaweeds are free floaters, growing unattached in vegetative islands. The mysterious Sargasso sea is filled with millions of tons of these free-floating kinds of sea plants of the *sargassum* group.

Rather than roots, stems, leaves, and seeds, seaweed plants have these main parts:

Holdfasts—instead of roots
Stipes   —instead of stems
Fronds   —instead of leaves
Spores   —instead of seeds

With this kind of botanical equipment, some kinds of sea plants can grow comfortably in 40 feet of water attached to rocks on the bottom and floating up to the top. To illustrate this, we include a drawing of the *Macrocystis* type of seaweed prevalent on the California coast.

50

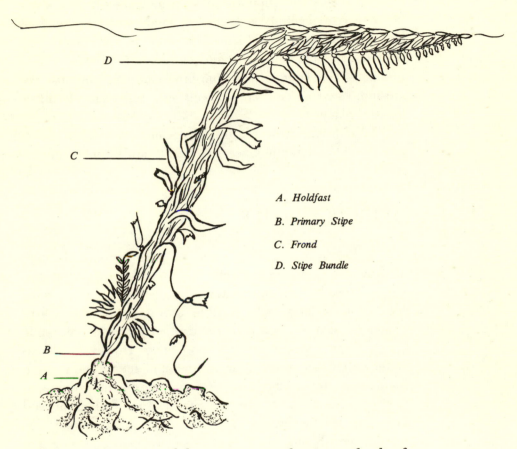

A. Holdfast

B. Primary Stipe

C. Frond

D. Stipe Bundle

Diagram of a young adult Macrocystis plant at a depth of about 30 feet.

Such a sea plant may produce 50,000 to 100,000 spores (seeds) to reproduce itself. Dr. Arthur C. Mathieson, director of the Jackson Estuarine Laboratory in New Hampshire says in "Seaweed Aquaculture" that some seaweed plants may produce as many as 1 million "swarmers" (spores), each capable of making a new plant.[1]

Since seawater supports these plants, they have a minimum of rigid lignin-type tissues of kinds found in stems of land plants. Therefore, seaweeds may be processed, extracted, liquefied and made into hundreds of kinds of useful products. This is important from our standpoint, because we are interested in *liquefied seaweed* as a supervalue fertilizer for land plants.

In addition, these features of ocean botany have special interest since they influence the quality and supplies of seaweed:

ALL SEAWATER IS FERTILE

Unlike land areas which may be arid or nonproductive, all seawater is rich in minerals and capable of growing seaweed. True, the mineral content varies. Cold ocean currents tend to be mineral-rich, capable of supporting more fish, planktons, diatoms, sea plants and other forms of life, while warm, tropical waters are less productive. However, all seawater contains all of the known mineral elements and is capable of growing seaweed. There is never a drought or crop failure in this universe of plants. Erosion of minerals from land areas into the sea steadily replenishes the food supplies. As we impoverish land by intensive use, we hasten erosion and enrich the sea.

SEAWEED GROWS IN COLD WEATHER

Recent studies of seaweed summarized by R. H. Mann show that northern varieties of seaweed are capable of growing rapidly even in winter weather. Mann says, "*Laminaria* and *Agarum* in eastern Canada perform the surprising feat of

[1] Arthur C. Mathieson, "Seaweed Aquaculture," *Marine Fishery Review* 37, 1975, National Oceanic & Atmospheric Administration, Washington, D.C.

growing rapidly throughout the winter when temperatures are close to 0° C. (32° F.) and light intensity is low. . . . Other species of perennial, subtidal seaweeds have been shown to grow throughout the winter." [2] This unique capability of sea plants to grow well in cold climates supports the idea that supplies will be adequate for horticultural and food uses.

SEAWEED IS CAPABLE OF HIGH YIELDS

Studies of seaweed at the University of California, summarized in *The Seaweed Story* (State of California, undated) show that a frond (leaf) of seaweed may grow 12 inches in a single day when sunlight and temperature are favorable. This may occur especially when water movements and tides bring fresh supplies of nitrogen to the plants.

R. H. Mann and his research group studied seaweed growth by punching holes in blades (fronds) of the plants and measuring the rate of movement of the holes to the tip, via growth. They found that the plants renewed their tissues several times a year. He says, ". . . the biomass of new tissue produced annually was up to 20 times the initial biomass of the blade. We were particularly surprised to find that growth in length was rapid throughout the winter, and that growth rate reached a peak in late winter and early spring when the water temperature was close to 0° C."

Commenting on the overall productivity of northern seaweeds, R. H. Mann said, "It is clear that seaweeds are among the most productive and that their productivity is as high as, or higher than, some of the most productive terrestrial systems."

WORLD HARVESTING AND PRODUCTION IS IN ITS INFANCY

With virtually limitless potential supplies of seaweed growing in vast fields of seawater, the present rate of harvest-

[2] R. H. Mann, "Seaweeds: Their Productivity and Strategy of Growth," *Science* 182, No. 4116, December 1973. Copyright 1973 by the American Association for the Advancement of Science.

TRY A KELP REMEDY

Centuries ago, old Norsemen learned to use seaweed as a remedy for cuts, scratches, abrasions, burns, stings, and nettles. They pounded the seaweed in a vessel until it was liquefied, and added wood ashes. The resulting gel was applied to the skin.

This remedy was quite sound. Potassium from the wood ashes provided a measure of antibacterial action, while the alkali extracted the alginate colloid from the kelp. Upon drying, this gel left an airtight film on the skin to prevent reinfection of the sore.

To make such a remedy at home today, you may use the following guides:

- For raw material, get a package of *laminarin* (kelp) powder from a health food store; or, some seaweed meal from a farm and garden store.
- Soak ½ cup of the meal overnight in a cup of water.
- Add 1 teaspoon of sodium bicarbonate (baking soda) and 1 teaspoon of sodium carbonate (washing soda).
- Transfer the mixture to a blender, and blend it at high speed for a couple of minutes.
- Pour it into a small jar. The remedy is ready to use.

For people near seashores: dulse, laver, Irish moss and other red seaweeds do not contain enough alginic acid to make the gel. However, any of the brown rockweeds, oarweeds or giant kelps will work nicely if you wish to gather fresh material.

ing and use is very small; comparable to cutting and use of forests by the Pilgrims in 1675 as contrasted with forestry and lumber production today.

W. A. Stephenson estimated world production at about 700,000 metric tons annually, with Japan accounting for over 50 percent of this total, and U.S. production estimated at only 2,000 metric tons a year.[3] However, California's alginate industry has increased our seaweed harvesting substantially since 1965 and it is now estimated at more than 100,000 metric tons a year (see "Marine Algae As An Economic Resource," Silverthorne, University of California; and Sorenson, Florida State University; unpublished paper).

As the needs for seaweed products increase and the uses expand, many other countries, including the United States, can follow Japan's example and grow, harvest, and process virtually unlimited supplies of seaweed, utilizing their adjacent oceans, coastlines, bays, and estuaries to produce and harvest this crop.

If America needs 1 million tons of seaweed a year, it can have 1 million tons. Our waters, including Alaska's fiords, can supply it.

SEAWEEDS OF ESPECIAL INTEREST FOR HORTICULTURE

Just as with land plants, seaweeds have evolved in hundreds of different species and families. Stephenson (cited above) says that over forty-two different seaweed families are found in British waters. The University of California studies (cited above) show that more than one thousand kinds grow along the Pacific coast, with about four hundred species in the Monterey area alone.

Based on evolution and ease of growth, these sea plants occupy favored water zones, just as land plants adjust themselves to different soils, elevations, and climates.

We are especially interested in four of these water zones, and four kinds of seaweed growing there:

[3] W. A. Stephenson, *Seaweed in Agriculture and Horticulture* (London: Faber & Faber, 1968).

1. **The Tidal Zone**
Ascophyllum nodosum

This is the home of the wrack-type seaweeds, such as the famous *bladder wrack*, characterized by having sacs or bubbles to float the plant in its briny place.

However, these seaweeds have the additional feature of growing in tidal waters, where they are alternately submerged in water and exposed to air with the movement of the ocean tides. Invariably, as young plants these seaweeds are attached to rocks or debris with holdfasts, only breaking loose in storms or at old age.

Ascophyllum nodosum illustrates this kind of seaweed. Growing along Scandinavian, British, and Irish shores, it was the easiest to harvest in large quantities during the past century; consequently, it has served as the resource for the Norwegian and British seaweed industries.

Hundreds of small enterprisers gather *ascophyllum nodosum* along Norwegian shores and deliver it to drying and processing places, where this seaweed is converted to seaweed meal, liquefied seaweed, and seaweed powder for use as feeds and fertilizers in many countries, including the United States.

2. **Mid-shore Zone**
Laminaria

Next beyond the tidal zone are waters of intermediate depth, fairly shallow but not depleted at low tides. This is the home of *laminaria* and other high production macro-seaweeds. Unlike *nodosum*, *laminaria* is submerged in seawater all the time, seldom exposed to air. As will be noticed in the mineral analyses of different seaweeds, this total submersion evidently increases the levels of some elements, notably iodine and iron.

Anchored to the bottom with holdfasts, *laminaria* grows abundantly at depths from 10 to 40 feet, or deeper in favorable places. Therefore, it lends itself to mechanical harvesting with giant mowing machines, such as those the Icelanders are using in their new seaweed industry.

Grown perennially under conservation-type management, this kind of seaweed is suitable for a large expansion in production to serve increasing world needs.

56

3. Deep-shore Zone
Macrocystis

Growing in deeper waters, this is the giant seaweed upon which California's seaweed and alginate industry is based. However, this sea plant is equally suitable for development to serve farm and garden needs.

Again, *macrocystis* uses a holdfast for anchorage to the bottom of its sea zone. It may thrive in water up to 100 or 120 feet deep, although 30 to 80 feet is a common depth for this plant. Alginate producers such as Kelco Company at San Diego harvest *macrocystis* with boats, mechanical cutters, and conveyors that deliver the crop onto the deck. It is then carried to a central processing plant for manufacture into various seaweed products.

This mechanized mode of harvesting will surely be used to gather seaweed for farm and garden uses.

4. Shoreline Zone
Eelgrass

Although not a true seaweed, this plant is included because many coastal Americans are interested in it. After publishing *Ecological Gardening for Home Foods,* we received many letters asking: "Is eelgrass good for garden fertilizer?" and "How can we compost and use eelgrass?"

Eelgrass grows in marshy locations, washed by ocean tides, accepting sea salts and ocean minerals in its life system. However, this is a flowering land plant with roots, sap, and a circulatory system. Somewhere along its evolutionary pathway it got ocean-minded and learned to live in seawater.

Although no research has been done on the values of eelgrass as a fertilizer, many gardeners have used it as a manure and source of minerals. It is surely a good addition to any compost pile. We believe that eelgrass can supply a full range of ocean minerals, but it will not have the colloidal and hormonal values of true seaweeds, or their ability to act as chelating agents.

If you have eelgrass supplies and seaweed is scarce, composted eelgrass can be used for mineral enrichment of your garden or small farm.

In addition to the sea plants described above, dozens of other seaweeds may be used as superfertilizers. Red, brown, and green seaweeds, large and small, are nourished by ocean water and have similar abilities to fertilize and protect land-based plants and crops. If tidal and near-to-shore kinds are not adequate for agricultural and food needs, the floating sea-weeds of the Sargasso sea can be harvested. Also, planned production of seaweed in suitable coastal waters can supplement supplies of wild seaweed as needs and demands increase for this versatile marine material.

GROWING TAME SEAWEED [4]

Seaweed cultivation in Japan began in 1736 in Toyko Bay, when seaweed food suppliers placed bamboo branches on the bottom and captured live spores. Carrying these branches to new areas of water, the Japanese workers were able to seed, grow, and harvest new supplies of an edible seaweed called *nori*.

Later, Japanese marine biologists found that the seaweed spores could be incubated in large numbers on clean oyster shells, then transferred in a "budding" stage to protected waters for rapid production of the seaweed crop. Use of this method with variations for different kinds of seaweed and growing areas has enabled the expansion of tame seaweed production in Japan into a large enterprise employing over 300,000 workers and yielding about 200,000 metric tons a year.

One of the interesting variations used by the Japanese in tame seaweed culture is called "rope" cultivation. With this method, rope or heavy twine is placed in the vicinity of "mother" plants to catch seaweed spores. These are incubated for several months until the spores are "budded," then attached to poles or rafts in selected seaweed growing areas. It is reported that a raft 6 × 12 feet in size may produce about 1 ton wet weight of seaweed per season in this rope and raft culture.

Feeding the seaweed for rapid growth has become part

---

[4] Part of the information in this section is based on *Seaweed Aquaculture*, by Arthur C. Mathieson, University of New Hampshire, published in *Marine Fisheries Review* 37, January, 1975.

of tame seaweed production. Earlier in the Japanese enterprise, this was accomplished in a mild fashion by placing the new plants in the mouths of estuaries and in currents of water known to be rich in ocean nutrients, especially nitrogen. However, in recent years the practice of fertilizing the crop is often used.

In Chinese production of *laminaria* variety of seaweed on rafts, fertilizer supplies may be provided. Commenting on this in "Seaweed Aquaculture" (cited above), Dr. Arthur C. Mathieson says, "A porous earthenware cylinder containing fertilizer is attached to each of the culture rafts. The fertilizer becomes dispersed in the water and it is absorbed by the sporophytes [young seaweed plants]."

The yield of cultivated seaweed varies, of course, with varieties and climate zones. However, it is indicated to be from 1 to 3 tons dry weight per acre per year. Chinese producers using a simple rock culture, placing seeded rocks on the bottom of the growing area, are reported to have attained a yield of about 1 ton per acre of *laminaria* seaweed.

In another measurement in Philippine waters, about 10 tons per hectare of dry weight per season was reported. This would amount to 2.5 tons per acre. At a value of $300 per ton, this would surely rival corn or cotton in value of crop as an economic enterprise.

PESTS AND OBSTACLES

Sea urchins are the arch-pests of seaweeds. They love to pasture on the tender fronds. When uncontrolled, the urchins may move in and destroy the seaweeds of an entire area. To cope with the sea urchins, Oriental seaweed growers may apply quicklime to the beds before planting to rid the area of these pests.

In addition, there is a tendency for onshore civilization to conflict with seaweed production. Oyster, lobster and other fisheries resist invasion of their waters by seaweed growers and harvesters. Also, pollution of the marine waters by sewage, industrial chemicals, and petroleum is a major hazard. If tame seaweed production becomes a major enterprise, the growing areas will have to be selected and protected with care. In this phase, it is very little different from land crops. Seaweed, too, should not be polluted or grown in the wrong places.

THE MINERALS AND MYSTERIES OF SEAWEED

To understand seaweed, it helps to be a mystic as well as a clear-thinking person, since seaweed is almost as complex as life itself.

You start with gene strips and enzymes, knowing that people and plants inherit their life structures and systems. The little engineers are enzymes. For example, heredity tells the baby-to-be, "Zap, build two eyes," and provides the blueprint for eyes. Then, enzymes select the nutrients and engineer the job.

To build and repair muscles, nerves, blood, glands, bones, and all the rest, the processes are the same throughout life—heredity patterns and enzymes govern the vital activities. So it is with plants, too.

In such a life system, minerals have an essential place, since every enzyme has a mineral or sulphur as a key unit in its chemical structure; and plants and people utilize dozens of different kinds of enzymes, each with iron, zinc, manganese, copper, cobalt, selenium, nickel, or some other mineral at its core.

In the slow process of exploring the unknown, scientists "discover," one by one, the different minerals essential for life and growth; and naturally they "discover" major needs first. For example, the importance of iron was learned early, since blood cannot be made without it; similarly, chlorophyll was identified in the case of plants.

After some one hundred years of research, the view prevails that about fourteen minerals are essential to nourish plants adequately, and about sixteen are needed by people.

However, this approach tends to overlook the possible biological roles of another thirty or forty minerals of our environment, including their possible roles in the complex enzyme and hormone systems of plants and people.

60

EDIBLE SEAWEEDS

Dulse, laver, and Irish moss are the edible seaweeds available in many health-food stores around the country. These are members of the red family of algae. In Japan, considerable use is made also of brown *laminaria*, its many products being called *kombus*. Sea lettuce, a species of green seaweed, has been collected and used fresh by coastal dwellers of many areas for centuries.

Dulse, the dried sun-bleached tissues of *Rhodymenia palmata*, is harvested from both northeastern and northwestern Atlantic shores. Ireland, Iceland, the Maritime Provinces of Canada and our New England coast are the plant's habitat. Dulse may be chewed when fresh or dried, or used as a chopped material added to soups and chowders. Simmered with milk, dulse was commonly eaten in Ireland.

More widely consumed is laver, made from various red seaweeds of the genus *Porphyra*. Dried or fresh, the preparation of laver consists of soaking it for an hour or two in water, chopping it, and simmering the pieces until they become soft. The water is poured away, and the tender residue may be mixed with oatmeal or cornmeal, formed into cakes and fried. Or, the laver may be seasoned, buttered, and served as a vegetable.

In Japan, this kind of seaweed is grown domestically, harvested, cut finely and dried in thin sheets. Called *hoshinori*, this food finds many uses in Japanese cuisine, including replacement of bread in rolled sandwiches. Laver and *hoshinori* have a substantial protein content, about 30 percent, as well as vitamin C and nutritious minerals.

With about 82 milligrams of vitamin C per 100 grams of weight, spring-harvested laver (*Porphyra*) is at least equal to citrus as a provider of this essential vitamin. Other edible seaweeds also have good vitamin C levels when harvested at the right time of year.

This is where seaweed comes in. Grown in earth's mineral-complete ocean water, it has in its tissues every mineral that *may* have a role in life systems, including those scientists will "discover" as being important fifty years from now.

An intuitive view is possible in assessing this nutritional situation; and it may prove to be valid. This is the contention. After emerging from ancient oceans in the first place, several eons ago, man's predecessors ate everything along their evolutionary pathways, and formed their organs and life systems with these full mineral nutrients. In that system-building process, it is probable that every active mineral found a role of some kind or other; and that equipping enzymes with key minerals was part of this evolutionary process.

If something like this occurred in the development of plants and people, it may be assumed that *all* biologically active minerals are useful to sustain life and health. There are about sixty such minerals, all contained in roughly sound assortments in seaweed. Laboratory analyses listing them are given in Appendix A.

When combined with fish to provide animal-based nutrients, seaweed is the most complete and useful mineral supplement available on earth to redress man-made deficiencies in horticulture—in both food and feed crops.

SEAWEED FOR CHELATES AND COLLOIDS

Next, let us remember that Nature did not offer iron to aboriginal man as iron ore or carpet tacks; or zinc to plants as sheet metal. Built into organic compounds of old foods, these minerals were chelated. This means that they were chemically protected within a molecule, and we ate the whole molecule, not simply the iron or zinc.

Derived from a Greek root, the term chelate means to hold as in a claw.

Seaweed holds its minerals in this way for use by plants and people, and it also may act as a chelating agent for other minerals in a feed or diet.

This value in seaweed is related to its companion feature of having a high content of colloids. Known also as gels and alginates, these portions of seaweed account for its uses in making soups, ice cream, agar, medical supplies, cosmetics, paints, inks, textiles, and dozens of other products.

Meantime, the colloids in seaweed are also valuable to condition and improve soils. If a soil is too sandy, seaweed gels will add substance and moisture-holding capacity. If the soil has too much clay, the seaweed will soften it and improve its ability to grow plants. That is why seaweed is often used as a soil conditioner.

HORMONES IN SEAWEED

However, minerals and colloidal properties are only the beginning of our interest in sea plants for superenergy in farming. Equally interesting and mysterious is the hormonal feature of seaweed.

To widen our shores of mystery and knowledge in this sector, let us start with an odd example of pheromones, which are used as communication tools between insects. For example, a female gypsy moth may manufacture and secrete small amounts of this substance. Molecules evaporate and, borne downwind, are sensed by male gypsy moths who home in on the scent and follow it to happiness. If insects can do this, utilizing pheromones as chemical messengers, can we deny that plants using their hormone materials and techniques might be able to attract insects one-quarter of a mile away, or that they might be able to *repel* insects?

To add to the reasons for being open-minded in this field, let us remember that aphids, using hormonal equipment, may have girl or boy babies at will. If they can do this, what are the limits imposed on hormones in plants as to pest and disease resistance, and as to achieving growth and yield increases without, necessarily, the need for more conventional NPK fertilizers?

In the case of seaweed, it is already well established that its tissues contain extraordinary amounts of plant hormones, and seaweed fertilizers and sprays may supply hormonal materials for use by other plants.

In *Seaweed in Agriculture and Horticulture* W. A. Ste-

63

phenson says, "Auxins (plant hormones) are substances which influence the activity of the living cells in plants by encouraging, and inhibiting, the extension, growth, and division of cells. They can be effective in concentrations of much less than one in a million." Stephenson then goes on to show (pages 90–94) that seaweed and seaweed products may provide auxins and gibberellins for use by other plants.

Dr. T. L. Senn and his research associates at Clemson University (Clemson, South Carolina) have demonstrated that seaweed meal and seaweed extracts contain substantial supplies of hormones which may beneficially influence growth and yields of plants fertilized with seaweed.

Commencing in 1959, Dr. Senn has demonstrated in step-by-step research that seaweed improves respiration and life in seeds and seedlings, and that timely applications promote growth and yields of plants in ways that cannot be explained by conventional horticultural theories and principles. In addition, he and his research group have established that foliar applications of seaweed to peach trees in June may substantially reduce molds and spoilage of the crops when harvested. And, along the way, they have noticed that seaweed materials may reduce insect populations on plants and assist some kinds of plants in resisting damage by insects and diseases. However, this is pioneer work in a large realm of the unknown, and the Clemson University seaweed research during 1959–1976 is mainly a fascinating invitation to many universities and private agencies to expand this investigation of energy-saving potentials of seaweed when it is used in farming and gardening. A summary of the Clemson work to date is given in chapter 9.

CYTOKININS

However, the most dramatic hormonal effects of seaweed are those attributed to *cytokinins*, reported by Dr. Gerald Blunden, research leader at Portsmouth Polytechnic Institute in England.

Cytokinins are a group of plant hormones similar to the auxin group. Although known by name in America, they have hardly been studied here as to their possible horticultural values. Certain seaweeds, notably the *laminaria* variety, contain substantial supplies of this hormonal substance.

In a report to the Eighth International Seaweed Symposium, August, 1974, Blunden said, "Seaweed extracts are characterized by their high cytokinetic activity. The most important effects of cytokinins are those on cell division, cell enlargement, the delaying of senescence and the related transport of nutrients."

Telling of spray applications of seaweed on sugar beets, Blunden said, "It is thought that . . . cytokinins are very restricted in their movement within the plant, if indeed they move at all from the original site of application. Treated areas act as metabolic "sinks," and amino acids, phosphates, and various other substances accumulate in the plant tissues directly under, or very close to, the site of application. More is involved than a simple mobilization of nutrients. . . . The observation that cytokinin treatment augmented the ratio of R.N.A. to D.N.A. suggested that a critical effect of cytokinins . . . might be the maintenance of protein-synthesizing machinery, perhaps by regulating R.N.A. synthesis. Hence, increases in leaf size, protein content, chlorophyll and leaf life would increase and prolong the photosynthetic power of the plant."

Treated with liquefied seaweed at the rate of only 1 gallon per acre, sugar beets in the English research plots showed a yield increase of about 8 percent attributed by Blunden to the cytokinin effect. Additional information on English seaweed research is given in the Appendix.

Now, getting back to insects smelling food or sex one-quarter of a mile away, it takes only small amounts of hormone substances to have large effects in the plant world. Very small applications of liquefied seaweed, providing cytokinins and auxin substances, may strongly influence growth and yields of plants.

Similarly, in their roles as enzyme activators, seaweed minerals in very small quantities may produce large effects on yields and quality of plants and crops. This is the genius of seaweed as an energy supplier from the sea: a little bit in farming and gardening may go a long way.

65

Recently the authors have observed yield increases of 20 percent to 50 percent in Midwest field crops from spray applications of only one pint per acre of "cultured" seaweed, diluted in water. Such results could not possibly be obtained via orthodox nutrition of the plants with nitrogen and other recognized nutrients. It has to be an enzyme and hormonal effect.

So, next time someone tells about growing corn 20 feet high or getting a bushel of potatoes from a single hill, just stay calm and say, "Yes, I heard about that. Hormones. Same thing that makes bees fly half a mile to get honey."

KINDS OF SEAWEED PRODUCTS

Seaweed in the ocean is useless to land people, and wet seaweed on the beach is expensive to ship into inland places. However, as we have mentioned, sea plants lend themselves to drying, processing, and liquefying. Consequently they may be shipped cheaply and used in many ways.

We learned a bit about this back in 1955–1960, when the main seaweed supplies for Oregon and Washington were called "Algit," shipped from Norway and England. First, we bought dry seaweed meal, then liquefied seaweed that looked like soy sauce from China.

Currently, seaweed may be obtained in the following forms:

**Wet Seaweed
on the Beach**

Just gather it and use it "wet." Washing off the salt is a good idea; composting is even better. Add seaweed up to 50 percent in the compost pile.

**Dry Seaweed
Meal**

The suppliers dehydrate and grind it into an attractive meal that smells like oceans and sea things. It may be applied straight at ½ to 1 pound per 100 square feet of soil area (250 to 500 lbs. per acre), or mixed with other fertilizers.

66

**Liquid
Seaweed
Concentrate**

This is the liquid form that looks like soy sauce. It is made by treating the wet seaweed with alkaline materials, and steam cooking it under pressure (autoclaving). Diluted with water, it may be sprayed on plants or applied to soils or seeds.

**Seaweed
Powder**

In this form, the seaweed is first liquefied, then reduced to a 100 percent soluble powder, similar to instant coffee. The farmer or gardener may then reconstitute it by dissolving it in water, subsequently spraying it on plants or soils.

**"Cultured"
Seaweed**

This is a new form in which seaweed is liquefied and treated to improve its efficiency as a fertilizer. In this process, magnesium, iron, or other minerals may be added to improve the balance of mineral nutrients.

**Seaweed
Blend with
Fish**

In this case, liquefied fish is blended with seaweed to form a *complete* fertilizer with substantial nitrogen and phosphate content. It is one of the finest fertilizer materials available, and will be discussed in later sections of this book.

In these various forms, farmers and gardeners may derive four separate and strategically useful values from use of seaweed:

1. To nourish and encourage soil bacteria that in turn capture nitrogen and release nutrients from soils to plants.
2. Directly to nourish plants with complete assortments of minerals and with nitrogen.
3. To support enzyme activity in plants.

4. To provide cytokinins and other plant hormones for use by land plants.

This is why we call seaweed a superfertilizer; an energy source to utilize in lieu of petroleum and natural gas. In chapter 6, after providing basic information about fish fertilizers, we shall give specific suggestions and guides for use of both of these energy-savers in farming and gardening.

KELP PICKLES

Do you feel bored with work and the monotony of life? Then cheer yourself up. Make some kelp pickles. Here is a good recipe.

For raw supplies, use bull kelp from the North Pacific, elk kelp from California, oarweeds (*laminaria* species) from Atlantic or Pacific, or any other large-growing brown seaweed.

Select sections from the stipes (stems) less than 1½ inches in diameter, peel them with a potato peeler, and cut them into ½-inch rings or disks. Soak them for 1 hour in ice water.

For 2 pounds of kelp prepared in this fashion, make a syrup by boiling together the following:

White vinegar	3 pints
Sugar	2 lbs.
Salt	4 tablespoons
Soy sauce	2 tablespoons
Curry powder (optional)	1 teaspoon
Stick cinnamon (in spice bag)	3 inches
Whole cloves (in spice bag)	1 tablespoon

Cook the drained kelp pieces in this syrup until they are tender. Remove them from the syrup with a slotted spoon and pack them in jars. Pour the hot syrup over them. Seal, and invert for a few minutes to sterilize the lids.

There you are! We could never have completed this book without a supply of kelp pickles.

5.

BLENDED SEAWEED AND FISH— THE PERFECT FERTILIZER

Lo! It turns out that Squanto, the celebrated Indian who put fish fertilizer under his corn plants, got that idea from England rather than from the ancient folk wisdom of his tribe. Europeans had been using fish to fertilize their crops for centuries. Squanto went over there and learned how to do it.

Documenting this, Lynn Ceci in "Fish Fertilizer: A Native American Practice?" shows that Squanto was kidnapped by Captain Thomas Hunt in 1614, and taken to Spain and then to England where he lived and worked for two years before returning to Cape Cod. "Thus," says Ceci, "in the years immediately preceding his appearance at Plymouth, Squanto had not resided in Indian settlements but in those of Europeans in both the Old and New Worlds. In European settlements, the use of fertilizers was a feature of farming technology since the Roman expansion if not earlier. The particular use of shellfish debris was famous in France since the medieval period, and the crops in the coastal zone . . . were so productive that the area was traditionally known as the 'gold coast.' Fish fertilizer was cited in an English publication in 1620, that is, before Squanto's appearance. Therefore, one cannot eliminate the possibility that settlers . . . already knew about the value of fish as fertilizer before the spring of 1621." [1]

[1] Lynn Ceci, "Fish Fertilizer: A Native American Practice?" *Science* (April, 1975). Copyright 1975 by the American Association for the Advancement of Science.

All of these people, including Squanto, discovered what we shall rediscover in America: fish is a gorgeous farm and garden fertilizer and mineralizer of food crops. Growing in seawater the same as seaweed, ocean fish are able to put into their tissues the same wide assortments of minor nutrients needed by plants, *plus* generous supplies of nitrogen and phosphorus not found amply in seaweed.

Therefore, as we shall discuss in this chapter, blends of seaweed and fish make excellent fertilizers.

VALUES OF FISH TO FERTILIZE PLANTS

As plucked from the sea, typical ocean fish contain about 20 percent protein. For example, Atlantic salmon contain 22.5 percent, halibut 19.8 percent, haddock 18.3 percent, shrimp 18.1 percent and perch 19.3 percent.[2] In turn, this protein is over one-sixth nitrogen, so that raw fish offer about 3 percent nitrogen to plants and crops.

When dried and processed as fish meal, the level of nitrogen rises to over 12 percent, since the water removed in drying contains no nitrogen; all is left in the fish meal. Therefore, a fertilizer-grade dry fish offers twice as much nitrogen as the 5-10-10 and 6-10-4 fertilizers sold in farm and garden stores.

However, these examples are of whole edible fish. How about fish-dressing wastes and hake, sharks, bullheads, and other "waste" fish counted as useless by commercial fishermen?

These materials, too, are premium-quality fertilizers. Heads, tails, fins, bones, and guts of fish contain the same kinds of major nutrients as the food portions, plus higher levels of minerals due to the inclusion of bones, blood, gland, and organ tissues in these wastes.

For every pound of edible fish sold in U.S. food markets, another pound is discarded as a dressing waste, and the discarded portion is a high protein-nitrogen material, suitable for making superior nitrogen fertilizers.

[2] From *Composition of Foods*. Agriculture Handbook No. 8. (Washington, D.C.: U.S. Department of Agriculture, 1963).

SUPERQUALITY PHOSPHATE

Next, energy savers should be aware that fish wastes and scrap fish have about 4 percent phosphate (P_2O_5) dry basis, of a superquality for farm and garden use. As constituents of bones and other skeletal structures, part of this phosphate is combined with calcium in a form similar to good old-fashioned bone meal.

In addition, however, part of this phosphate is organically bound and soluble—in effect, chelated. This superior fish phosphate does not "tie up" in the soil, is easily found and used by plant roots, and is truly a "super" form of phosphate for farm and garden use.

OTHER MINERALS

Fish fertilizers resemble seaweed in that both have 3 percent to 4 percent potash (K_2O), dry basis, and both contain large assortments of other minerals such as iron, zinc, manganese, copper, magnesium, and iodine. However, in the case of seaweed nourished directly from seawater, it takes in and concentrates large amounts of sodium and chloride—sea salt—while fish have mechanisms to avoid this.

The gills of marine fish have salt-secreting tissues which assist them in returning to the water excesses of sodium and chloride which they may have assimilated in foods. Also, these mechanisms in fish enable them to reject excesses of aluminum and strontium which are freely taken in by seaweed, but are undesirable in fertilizers for land plants, due to their interference with iron and calcium nutrition.

Let us summarize in this fashion, comparing fertilizer values in fish and seaweed:

- **Nitrogen.** Fish provides abundant supplies, 12 percent to 15 percent dry basis, while seaweed offers less than 1 percent.
- **Phosphate.** Fish provides about 4 percent dry basis, of a superior form of phosphate, while seaweed offers less than 1 percent.
- **Potash.** Both are good potash sources, with 3 percent

to 4 percent of this mineral. However, a high sodium chloride level in the seaweed may reduce its ability to transfer the potash to plants.

- **Other minerals.** Seaweed provides the greater assortments. Fish has fewer minerals but in better balance.
- **Chelates.** The minerals in fish are mildly chelated, being built into the protein structures. In seaweed, the minerals are strongly chelated and seaweed has the ability to act as a chelating agent within plants to which it is applied . . . a helpful feature.
- **Hormones.** Seaweed is a superior provider of useful hormones, notably *cytokinins* and auxins, while fish have little or no capabilities in this field.
- **Colloids.** With abundant colloids, seaweed is a gorgeous soil conditioner, while fish does not claim an important role in this sector.

By putting these values and attributes together, we find that blends of seaweed and fish are logical and of especial interest as we *increase* food production in the United States, while at the same time *reduce* our demands on gas and oil. Seaweed and fish can serve as strategic energy-savers in farming and gardening.

FISH FERTILIZERS IN 1877— A HUNDRED YEARS AGO

As Joe Garagiola said about selling cars in the Great Depression of 1975, "We did it once, and, by crackey, we'll do it again." The same may be true for making and using fish fertilizers in the United States.

Back in 1877, a hundred years ago, Luther Maddocks of Boothbay, Maine, was America's first seaweed businessman; also, he was a leader in our country's first large-scale fish fertilizer enterprise, making and selling over 100,000 tons during 1873–1877. Typical of many earlier Americans, Maddocks was an active community builder and also a keen observer and regional statesman.

Listen to Luther Maddocks in his brief book *The Menhaden Fishery of Maine:*

74

The ocean is the grand receptacle in which the fertility of the lands continually hastens to bury itself. So long as a land surface is uplifted above the sea, the rains wash the finer parts of the soil, and especially its soluble and chief fertilizing parts, into the streams, by which they are ultimately deposited in the ocean. The great cities of the world, nearly all of which lie upon coasts, or upon streams communicating with the sea, and towns and villages innumerable, have been for long periods, and still are giving up the chief part of the waste and rejected material of food to the same grand depository. Over a hundred million dollars' worth of provisions change hands in Washington market of New York City each year. In addition, two million barrels of flour and a vast amount of other grain are consumed in the same city and its immediate environs; besides groceries and other food supplies. The city of London is estimated to use four million barrels of flour yearly, and over 500,000 cattle, sheep, and swine are sold at the Metropolitan market alone. Poultry, game fish, eggs, etc., are consumed by millions. The residuum of this great consumption is to a large extent lost in the ocean, and these are but two examples.

These immense contributions, once received in the sea, are distributed in various forms. Some of it lies dormant in mud flats. Some is in the beds and bodies of shellfish. Much is absorbed by marine plants, which are consumed by marine animals, to be devoured by other marine animals. Thus the abundance of the sea is increased at the expense of the land. No systematic efforts on a considerable scale have, until recently, been made to reclaim any considerable part of this waste.[3]

Luther Maddocks then goes on to show that Menhaden fish scrap contains about 4 lbs. of ammonia per barrel, plus

[3] Luther Maddocks, *The Menhaden Fishery of Maine* (privately published 1877), State Library of Maine.

3.2 lbs. of phosphoric acid. And "when dried so as to reduce the water to about 20 percent or 25 percent, it can be, as before observed, ground to a powder and transported to any distance. In this state it yields about 9 percent of ammonia."

Maddocks and his business associates in Menhaden Oil and Guano Manufacturers of Boothbay, Maine, made such a fish fertilizer material. He says:

> Nine-tenths of the scrap turned out at the works of the Maine Association are bought by the manufacturers of super-phosphate to ammoniate their products, of which 400,000 tons are produced yearly in the United States. They combine it, when dried and pulverized, with South Carolina phosphatic rock, ground bones, with imported guano deficient in ammonia, etc. It is understood that not over one ton of the fish guano is used in connection with three or four tons of the mineral ingredients. They doubtless use as little as will yield the proper percentage of ammonia, since the cost of drying and other treatment, with the labor involved and great reduction in weight, makes it a dearer ingredient than bones. Ammoniated superphosphate sells at from $35 to $45 per ton. When honestly made it is known as a strong and durable fertilizer, adapted to general use, especially for hoed crops; but for grass, which is the staple crop of Maine, it is considered by all agricultural chemists that a fertilizer containing a larger proportion of nitrogen, with a smaller content of the phosphoric element, will give better and more results.[4]

This kind of technology to recover nitrogen and phosphorus in fish wastes is the kind we may need to reexamine and employ a hundred years later in 1977, as gas and oil get too scarce and too high priced to be used in making fertilizers and pesticides.

Moreover, we'd gamble that the old superphosphate of

[4] See previous citation.

1877, made with fish, Carolina rock, and bones was a great product. We should love to have some for our own gardens.

WHAT IS FISH "EMULSION"?

If you ask one hundred home gardeners if they ever heard of fish fertilizer, many of them will reply, "Sure, I've used fish emulsion." They are referring to a fairly smelly product sold under the name "emulsion" in farm and garden stores.

On the market for many years, this fish material is not an "emulsion," nor is it representative of versatile liquid fertilizers that can be made from fish wastes and whole waste fish such as hake, sharks, bullheads and surplus carp.

Fish "emulsion" is a by-product of the fish meal industry, made from the so-called stick water that remains after removal of solids and oils. The solids become fish meal sold for poultry and animal feeds; the oils go to oil products manufacturers. The remaining broth is the material from which fish "emulsion" is made.

At an earlier time, farmers tried to use this residual liquor, or stick water, without further processing, as a liquid fertilizer. However, the stuff was too unstable, fermenting like wild home brew. Currently, these fish solubles are evaporated to about 50 percent solids, making a rather thick, goopy end product that is bottled and sold as fish "emulsion."

Since the oil has been removed, this fish product can hardly be called an emulsion. It is more accurately called fish solubles, being the nonoil and nonsolid portions of the fish.

As sold in garden and nursery markets this kind of fish fertilizer contains about 5 percent nitrogen, 1 percent phosphate and 1 percent potash, a good fertilizer but a bit low in plant foods and rather difficult to handle and use.

VARIOUS KINDS OF FISH-BASED FERTILIZERS

Back in 1950–1960, the authors probed the rich potentials of fish fertilizers, using fishery wastes of coastal Oregon

77

GOURMET COWS AND SNIFFY RABBITS

In 1954, sitting on a grassy knoll in Skagit County, Washington, we watched a herd of cows grazing in a 40-acre grass pasture. But it was an unusual pasture: half of it was fertilized with a mineral-rich *complete* fertilizer and the other half with regular old NPK; and a stream ran alongside the NPK half, opposite from the mineral-rich part, like this:

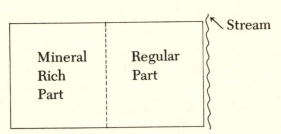

As is their wont, the cows grazed a while then went for a drink of water. However, in a week's watching we never saw them take a bite of grass from the unmineralized portion, although it bordered the water supply and they had to walk 200 yards to return to the grass of their choice.

The gourmet cows knew the invisible boundary of the well-fertilized grass in this high rainfall country. As the camels used to say, "We'd walk a mile for some good grass."

A few miles to the north, some rabbits showed a similar intelligence about their food. In this case, a farmer planted a field of strawberries using the usual practice of fertilizing the new plants with dilute phosphoric acid to prevent transplanting shock and start them on their way. However, we gave him a gallon of liquid fish fertilizer to try on a few rows for comparison with his regular transplanting material and these rows were staked and marked.

A month later the farmer called and said, "Come out and look at the strawberry field." We did, and this is what we saw: The "regular" part was doing nicely, but damn! the staked rows were almost bare of leaves. A community of rabbits from a nearby woodlot had grazed these "sweet" plants nearly to the ground without touching any other strawberry leaves in the whole field.

Gourmet cows and sniffy rabbits. . . . Do *you* think a smart rabbit would tarry long in the produce section of a supermarket, or that a self-respecting cow would eat Wheaties?

We wonder how people would act if they might use ancient wisdom choosing their foods.

and Washington. We bought many truckloads of crab and shrimp meal for use in making special fertilizers for fruit growers, nurserymen, and gardeners of that area. This work was done in commercial operations of Chas. H. Lilly Co., Pacific Agro Co., and Alaska Agro Co.

By providing markets for producers of organic waste materials, we were able to stimulate and support small business units in waste processing, such as Frost Fertilizer Co. and Helms Feather Meal Co. In the case of Helms, we assisted him in successfully hydrolyzing poultry feathers and wastes, making a 12 percent nitrogen product called feather meal.

On the docks of Bellingham, Washington, Dick Simmons adapted a similar technology for hydrolyzing fish dressing wastes from the Bellingham fishery whose boats ranged from Alaska to California. Using heads, tails, viscera, and whole scrap fish, Dick made high analysis liquid fertilizers, for example, Hi-Tide 10-5-5, as a translucent, odorless, easy-to-handle product. Timely again, this technology is now being reintroduced at Carpole's Inc. in Minnesota (see chapter 6).

Based on 25 years' experience using and processing fish wastes, we see the following kinds of fish fertilizers as having useful roles in energy-saving and cost-saving modes of farming and gardening.

1. Fertilizer-Grade Fish Meal

The purpose of this material is to stimulate and encourage soil bacteria, as well as to nourish plants.

In recent years, regular fish meal has become almost exclusively a feed material for poultry and livestock, too precious to use as fertilizer. That may change as fish wastes are made into a fertilizer-grade fish meal, to recover nitrogen and other nutrients.

This kind of fish meal product can be made by liming, composting, and drying fish wastes along with other useful materials, somewhat like Luther Maddocks did with his fish scrap, rock phosphate, and bones back in 1877. Such a local enterprise can be conducted whenever crab or fish wastes accumulate. It can be used to process whole fish that commercial fishermen are now throwing back into the water as nonsalable kinds, as well as fish-dressing wastes.

In the case of crab materials, only 15 percent of the weight of crabs is used as meat; the other 85 percent is discarded and often becomes a troublesome pollutant. However, the crab-dressing wastes have excellent fertilizer values, with about 6 percent nitrogen (dry basis), 2 percent high-quality phosphate, 2 percent potash and a generous provision of lime from the crab shells.

All of these fishy wastes are eligible for use in making fertilizer-grade fish meal in local enterprises, wherever fish are landed and processed.

Blended with seaweed meal, these products are superfertilizers for use in rebuilding bacterial populations in soils. In Northwest horticulture we have seen similar products generate yield increases of 20 percent when applied at only 200 or 300 lbs. per acre, because they nourished bacteria that in turn captured nitrogen from the air and liberated mineral nutrients from otherwise inert soils.

This is the secret of energy-saving horticulture: build a fertility base in the soil with strategic materials such as seaweed and fish, so the soil bacteria, in turn, will fertilize the crop.

2. Liquid Fish Fertilizers

As noted above, certain constituents of fish are already water soluble and may rather easily be made into liquid fertilizer. These constituents include gelatin tissues, nonprotein nitrogenous units, protomins, carbohydrates, and water-soluble minerals.

The total of these makes up old-fashioned fish "emulsion," the by-product of the fish meal industry, which is not an emulsion but rather the fish solubles.

However, our aim is to utilize whole fish wastes and this requires solubilizing the nonsoluble tissues. It is done by a process called hydrolysis, used by Dick Simmons on the Bellingham docks in 1950–1965, and currently by Carpole's in Minnesota, as we shall describe in chapter 6. It produces a stable, attractive, easy-to-use product.

In Carpole's operation, a typical product contains 8 percent *N*-itrogen, 4 percent *P*-hosphorus and 4 percent *K*-potash plus other mineral nutrients of the fish. When made at this NPK level, each gallon of liquid fertilizer contains about 6 lbs. of actual raw fish.

When combined with liquefied seaweed at a rate of 75 percent fish to 25 percent seaweed, this becomes a 6-3-3 grade seaweed-fish blend; one of the most interesting and useful fertilizer products available in America today.

FUTURE FISH FERTILIZERS—
THE "ORGANIFORM" PROCESS

In previous pages we have probed only lightly the potentials in making nitrogenous fertilizers from fish wastes. To illustrate other methods, we shall briefly describe the "organiform" process developed by Dr. James O'Donnell, President of Organics, Inc., Slatersville, Rhode Island. O'Donnell has been a leading scientist in the field of nitrogen fertilizers since 1954, when he assisted Dr. K. G. Clark of USDA, and others, in the successful development of ureaform, the slow-release nitrogen of turf and nursery fertilizers.

In the "organiform" technology, any major organic waste may be reacted with liquid ureaform to upgrade the waste and make it into a high-value fertilizer material, capable of being shipped and marketed economically throughout a wide area. For example, the municipal wastes and sludge of Winston-Salem, North Carolina, are treated in this fashion to make a 20 percent nitrogen, clean, pelleted fertilizer; and leather wastes of New England are converted into "organiform" at Slatersville, Rhode Island, using the same versatile process.

Experimental work has been done at the University of Rhode Island adapting the "organiform" process to handle both liquid and solid fishery wastes—the liquid effluents and sludges as well as solid portions of fish.

The outlook is good for use of this process wherever large quantities of fish wastes accumulate and need recycling into protein and nitrogen markets—into feed and fertilizer uses.

Dr. O'Donnell says, "In the case of fish wastes we have a favorable situation for use of the 'organiform' process, because these wastes have a high intrinsic value. I see no serious obstacles to converting these high protein materials into super-quality fertilizers, solid or liquid."

As the cost of fertilizer nitrogen rises above 20 cents per pound, due to increasing scarcity and cost of natural gas, we foresee a scrambling for practical ways to recover and use the nitrogen now wasted in fish. At that time, we shall hear more about the "organiform" process, since it lends itself to use in large-scale processing of major organic wastes of the cities, farms and fisheries of America.

FOUR KEYS TO LOWER FOOD COSTS

When it takes 1,500 lbs. of purchased fertilizer per acre to grow food crops, at $100 per ton, that is $75; plus $25 for pest control, making a total of $100 cash costs in this department.

As costs of gas and oil go up, these farming costs will soar upward, too, unless cheaper ways are found to fertilize and protect the crops.

In summary, fish and seaweed, separately and blended together, liquid and dry, may assist in four ways to reduce needs for purchased fertilizers and pesticides, thereby saving energy and lowering food costs. We shall enumerate these ways.

1. Mobilizing
Soil Bacteria

Fish and seaweed provide whole cafeterias of mineral foods for these little workers, so that they may capture nitrogen from the air and release soil-bound nutrients, supplying nonpurchased fertilizers for the food crops, thereby reducing costs.

On many soils only 200 to 300 lbs. per acre of fertilizer-grade fish and seaweed meal will assist in this process of mobilizing soil bacteria.

2. Spray-Feeding the Crops

Liquid fertilizers sprayed on foliage are from 200 percent to 400 percent more efficient than soil-applied fertilizers for many kinds of crops and soils. Liquid seaweed and fish materials take advantage of this, raising efficiency in the use of limited nitrogen and mineral fertilizer supplies, thereby reducing costs.

3. Reducing Needs for Pesticides

Full mineral nutrition of the crops with seaweed and fish improves resistance to insects and plant diseases, and mobilizes enzymes and hormones of plants in defense against pests. This may reduce pesticide needs and costs by as much as 80 percent, while also improving safety and nutritional values in the foods.

4. Providing Plant Hormones

Liquefied seaweed provides cytokinins which in turn improve photosynthesis and the use of carbohydrates in plants and crops, extending the growth and fruiting season. Also, seaweed provides auxins and gibberellins which are useful for some kinds of crops. As a result, high yields may be obtained with less need for purchased fertilizers, thereby conserving energy and reducing costs.

Frequently we have been asked this question: "If these strategic materials are used, how much can the amount of conventional NPK fertilizers be reduced while still getting equally high yields of crops?"

Unfortunately, no one can answer this question factually and with research evidence, since federal and state agencies have done very little work in this field.

However, as practical agronomists we estimate that 25 percent of the petro-based fertilizers and pesticides can be eliminated within five years without reduction in crop yields

84

or national food supplies if these energy-saving materials are intelligently used. About 50 percent of the petro-based products can be eliminated within ten years through these and other changes in farming practices, without curtailment of crop yields.

Eventually, *all* petrochemicals can be eliminated, and will be—because the world will nearly run out of gas and oil by the year 2025. Much earlier, however, none will be used to make fertilizers and pesticides, because gas and oil will cost too much. Instead, we shall use seaweed, fish, sewage, manure, humates, compost, snake oil, whiffle-dust—everything containing nitrogen. We shall use them all because we have to do so or starve to death.

POWDER BLUE NITROGEN

In 1956, Jim O'Donnell was making "blue chip" Nitroform percent 38, the thinking man's fertilizer, which released its foods gently with the help of temperature, moisture, and soil bacteria. It looked like clean blue soap chips.

The by-product was "powder blue," a nitrogen-laden dust formed while making "blue chips."

Our man Bill Turney said, "Powder blue is insoluble in water, but it will break down slowly on the leaves of plants. Let's suspend it in water, with a spreader-sticker, and feed some apple trees with it."

On Garretson's big apple orchards near Yakima, Washington, we did that very thing. Using a regular spray rig, we painted a whole block of trees blue. We fed them that way, instead of with ammonium nitrate, the usual apple growers' fertilizer of that time. Then we added fritted trace elements (FTE) to part of the mixture, and let that slow-soluble material seep into the leaves, too.

Admittedly, we were twenty years ahead of the times. Neighbors said, "What are you saving fertilizer for? The Valley is full of it. Cheap."

However, fascinated with slow-release fertilizers that act like humus, we did another crazy thing. Chrysanthemum and Easter lily growers wanted anything that would add quality to their crops and control budding and blooming time.

So, Bill Turney and Dick Simmons said, "We'll make a miracle for you." They had a plumber make an 8-inch cylinder 4 feet long with a ¾-inch pipe coupling on each end. Then they cast "powder blue" Nitroform (ureaform) slugs 8 inches by 4 feet, to fit into the cylinder, but with a hole lengthwise so irrigation water would flow through, eroding bits of nitrogen, thereby gently feeding the plants.

Did it work? Of course, it worked. But remember, we were twenty years ahead of the times.

6.

CARPOLE'S—A BIG FISH STORY

Much older than people, the ubiquitous carp goes wherever we go; annexing lakes and streams; rooting, grazing, schooling, migrating; the smartest fish that ever swam. Ask any Minnesotan forty years ago, "What kind of fish in these cold lovely lakes?", and he would reply, "Pike, bass, muskies, perch, trout . . . greatest sport fishing in the world."

Today, carp, millions of them, threaten to destroy the multimillion dollar recreation and commercial fishing industries of Minnesota, Michigan, Wisconsin, the Dakotas and all other lake states. Carpole and his fishermen caught as many as forty tons of carp from a single drag of their seine while harvesting carp from Minnesota lakes in winter 1975–1976. Many of these sports fishing lakes are producing 500 lbs. of carp per acre per year, over 100 tons per season in a 500-acre lake; and unless they are removed, the days of sports fishing are numbered. Carp are taking over the fish pastures of mid-America.

CARP

In the annals of folk wisdom is the advice to eat the brain of a carp and thus become smart enough and strong enough to

89

overcome. That advice is based on observations that this old fish has learned to thrive any place where people live and pollute the water.

Fossil imprints show that carp were well established 2-million years ago, moving south in the ice ages and north again as the ice retreated. Originating in Asia, they followed people into the Mediterranean region. Romans and monks valued carp for food and took them into western and northern Europe.

Carp first entered the United States via California, where they were brought by German settlers around 1870. Also, this fish was introduced in eastern American waters by immigrants who prized carp meat as food. Its reputation as a food fish is a good one. Howard T. Walden in *Familiar Freshwater Fishes of America* reports that the Ostrogoths around 500 A.D. took steps to improve supplies of carp for the king's table. And, he says, "Culture of the carp in well-managed ponds has made this fish an important source of food in Europe and Asia for centuries. Fancy strains of carp have long been bred in central Europe for gourmet markets." [1]

There is little doubt that carp flesh is edible, that this is potentially an important food fish in America as our needs for low-cost protein foods rise, along with an increase in population. However, sports fishermen and conservationists usually hate carp because they muddy up clean waters, destroy spawning grounds of other fish, and eventually take over the waters if permitted to do so.

Carp are omniverous. They may eat bugs, worms, frogs, snails, eggs, and other fish, or pasture on water grasses and vegetation. Moving across a lake, they root like hogs, raising clouds of mud whetever they go. Mature carp in nutritious waters may reach 3½ feet in length and weigh over 50 lbs.

Minnesota's record for a carp caught on rod and reel is a 42-inch giant weighing 55 lbs. caught by F. J. Ledwein in Clearwater Lake in 1952. The world record is one caught in South Africa weighing 83 lbs.

As these reports indicate, carp are a versatile, enormously productive kind of fish, copiously present in inland

[1] Howard T. Walden, *Familiar Freshwater Fishes of America* (New York: Harper & Row, 1964).

waters of America, and if useful for fertilizer, surely in ample supply.

OLE CARPOLE

Ole Carpole is Maynard Olson of Garfield, Minnesota, near Alexandria. All his friends, admirers, and detractors call him "Ole." One thing for sure: if you meet Ole you may never forget him. If you are in his way, he may run right over you. So far, nothing has stopped Ole Carpole.

He got the name by first hating carp more than anyone else, then deciding to catch them and make them into foods and fertilizers. The Minnesotans called him Carp "Ole," so he named the business Carpole's, Inc., Minnesota's first big-time fish food and fertilizer factory.

Widely experienced and deeply informed, Maynard Olson is no fish freak or accident in Minnesota affairs. As a Construction Sergeant in the Korean War, he supervised several hundred workmen building and maintaining airports in Asia. Consequently, he has a fair knowledge of how to build and operate various things in many places.

Ole's experience with carp commenced in 1955, when he helped state game wardens to install and operate carp traps in the Alexandria area. After serving for fifteen years as a Director of the Viking Sportsmen's Association, with over four thousand members, Ole became vice present of that organization, which has served as a supporting base for the business of harvesting carp and converting them into foods and fertilizers.

CATCHING CARP NO. 1—
SEINING UNDER THE ICE

Seining under the ice is one of the principal ways to catch carp and other unwanted fish in mass production quantities. Using fish traps is the other way. First, we shall describe seining. This is how it is done.

In this example, Adolph, Tony, and Dick Piechowski, hearty Poles of northern Minesota, are the skilled leaders of

91

the project. They know how carp think, flock, and hibernate. Their equipment consists of:

- a seine 3,000 feet long and 25 feet deep, with a big "pocket" in the middle
- ropes, tractors, big fish boxes and gear
- tractor-mounted augers to bore 18-inch holes in the ice
- long wooden "needles," made of 1" x 4" lumber, to thread the seine from hole to hole under the ice
- sonar equipment, to locate exactly where the fish are lying under the ice

First, the Piechowskis get a permit from the State Fish and Game Commission to harvest the fish from a particular lake with state supervision, and the operation is scheduled; for example, on Friday, January 16. Then, a crew of twelve to fifteen local people is employed at about twenty dollars per day each to help catch the fish.

Headed by Dick Piechowski, this aggregation gets out of bed at 4:00 A.M. on the cold January morning and assembles at the lake.

On one side of the lake, they dig an "in hole" about 10 x 15 feet for use in putting the big seine under the ice. Then they bore auger holes about 50 feet apart and "thread" ropes across the lake, using the 1 × 4 wooden "needles" to do this job. Thus, the seine is put in place, under the ice, with ropes attached for pulling it.

At the other side of the lake, Piechowski has in the meantime dug a big "out hole," from which the fish will be taken. The ropes, of course, are threaded to the "out hole"; and somewhere between the "in hole" and the "out hole" are a whole lot of fish.

With the ropes pulling easily at first, the drag begins; then as the seine fills with fish, tractors are employed to haul them in. Sometimes the catch is so large it jams the area by the "out hole" and special work must be done to unscramble all those fish.

At the "out hole," a skilled workman uses a long handled "dipper," 6 feet in diameter, to dip fish from the seine, and put them on a sorting platform. Wiggling madly, the fish land there—carp, trout, muskies, perch, bullheads, whitefish—all kinds of fish.

State game wardens supervise the operation as the fish are sorted into the various boxes. The game fish go back into the lake, or into another one nearby. The carp and other "waste" fish are loaded on trucks and hauled to Carpole's for processing into minced fish meat and fish fertilizer.

How many fish are harvested in one haul? Often, 80,000 to 100,000 lbs. of the carp and other "waste" fish, which may be increasing at over 500 lbs. per acre per year in some of these find Minnesota lakes.

Carpole pays from 1 cent to 5 cents per pound for the fish. His payments reimburse the Piechowskis and their crew, creating new employment and cash flow in these job-scarce communities. So far, federal and state agencies have not assisted enough in this worthy enterprise, which is so skilled, so sound, and so hard to develop.

CATCHING CARP NO. 2—IN FISH TRAPS

The carp, smart old fish, get frisky in springtime, flock together, migrate, and spawn in the last of May or early June. Since the lakes of this region are interconnected by flowing streams, this means a great bunch of carp may rush suddenly from one lake to another. Sensing the weather, they like to do this just before a spring thunderstorm.

The Viking Sportsmen Association and the State of Minnesota are taking advantage of this migration habit of carp to catch them in fish traps located in or near the streams connecting various lakes.

A typical trap is simply a concrete structure into which the fish must swim, but from which they cannot escape; whereupon they may be dipped into boxes and hauled to Carpole's for processing.

In 1975, a consortium of federal and state agencies including the State Natural Resources Department and the Soil Conservation Service appropriated about 3 million dollars to build eighty-seven fish traps in Douglas county, Minnesota, as a pilot project in carp control for the region. This array of

93

traps will serve as a catching facility for Carpole's and any other similar enterprises which convert carp into fish meat and fertilizer.

QUANTITIES OF FISH AVAILABLE

Naturally, two questions arise: "How big is this waste fish resource?" "Is it really big enough to serve as the basis for a major food and fertilizer business?"

Part of the answer is given in the information for Douglas county where these eighty-seven fish traps will be installed for catching carp.

With over one hundred sport fishing lakes, Douglas County has approximately 6,000 acres of productive lake area, yielding about 500 lbs. of carp and other waste fish per acre per year. Even at two-thirds that rate, this amounts to about 1,000 tons of fish available for processing—enough to make over 1 million pounds of high protein fish meat and 200,000 gallons of liquid fish fertilizer annually.

It is a 2 million dollar local enterprise capable of creating and sustaining thirty to forty jobs catching, processing, and marketing carp—the world's most loved and hated fish. Remember, goldfish are of the carp family, too.

Projected into regional magnitudes, *if* Douglas County, Minnesota, equals 1/100 of the Lake Region's carp waters, then the total fish fertilizer resources of mid-America are:

- an annual yield of about 100,000 tons of carp and other waste fish materials
- containing about 6 million lbs. of actual nitrogen
- which, if made from gas, would require about 100 million cubic feet of natural gas to manufacture
- sufficient to make about 20 million gallons of 8-4-4 grade liquefied fish *

This oversimplified calculation gives an inkling of the fish resources that may exist in inland America. And remember, if

---

* In Carpole's process of hydrolizing fish, some nonfish nitrogen is used.

the Lake States do not provide these quantities of carp and other fish wastes, the waters of rivers and lakes further south will add to the supplies. Eventually, the dressing wastes from many catfish ponds will make their contributions, too.

CARPOLE'S INC.—THE BUSINESS UNIT

Incorporated under the laws of Minnesota, this unique business unit has over two hundred and fifty shareholders, many of whom are members of the Viking Sportsmen Association. They have provided about $200,000 with which to build this enterprise in foods and fertilizers. Maynard "Ole" Olson runs it, Ms. "Lannie" Olson is the check-writing comptroller, Fritz Knauer manages the plant, the Zeithamer family and other capable neighbors serve on the Board of Directors. This is a tight, well-run ship, catching and processing mid-America fish.

The plant equipment for producing the fish meat is of the Japanese type, capable of handling about 3,000 lbs. of fish per hour, separating meat from bones, yielding about 50 percent of the gross weight as a minced fish product. Under Fritz's management, this phase of Carpole's can produce 20,000 lbs. of carp meat per day.

The minced fish, 20 percent protein, can be made into fish sticks, patties, "hamburger," meat loaves, sausages, marinated "herring," fish stew and a dozen other commercially valuable foods. Carpole's fish is also seen as a low-cost, high-protein food for hospitals, rest homes, state and federal institutions, school lunches and military services.

And still, there is 50 percent of the fish left as waste for making fertilizers.

CARPOLE'S FISH FERTILIZERS

In the Carpole technology provided and monitored by Dick Simmons, the "wet" fish wastes and whole fish are hydrolized to break up the tissues and liberate their protein and other fertilizer nutrients. This is done in a steam-cooking pro-

AMERICA NEEDS LOW-COST PROTEIN
AND CARPOLE'S CAN PROVIDE IT

With 25 percent fat, regular-grade supermarket hamburger has about 15 percent protein, or 2.4 ounces of protein per pound. At 96 cents a pound for the hamburger, and disregarding other nutrients in it, this comes to about 40 cents per ounce for beef protein.

In comparison, carp meat is 18 percent protein,[2] and has a lower inherent cost than grain-fed beef, requiring no fertilizer or land costs and very little labor in its production. If and when the Food and Drug Administration and USDA permit its sale in food markets, carp protein will cost *less than* 40 cents per ounce. Furthermore, as a fish food, it may have nutritional values superior to beef.

Already, Fritz Knauer, Carpole's old-country meat master, is making the supercarp from Minnesota's cold lake waters into delicious fish patties and gorgeous-tasting fish sausage—gourmet items that would command three dollars per pound in European markets.

School-lunch kitchens, day-care nurseries, homes for elderly people, hospitals, state institutions, and military services need these low-cost high-nutrition foods. Carpole's is ready today to fill orders for one million two-ounce fish patties for school-lunch programs, or a mile or two of delicious fish sausage.

But no one at Carpole's is nervous about these markets for the valuable fish meat. Maynard Olson says, "This year we're developing our skills and large-scale supplies of fish and we are very busy making liquid fish fertilizers. Next year, we'll think about selling high quality fish meat, and I don't worry about the markets. When hamburger gets up around two dollars a pound—and it's going there—don't worry, the meat buyers will walk up here if they can't get gas to buy these fish products. I feel great. I eat a pound of carp a day just to keep in good condition."

Indeed, this is a timely new business enterprise, converting the Lake States cold-water carp into low-cost nutritious foods.

---

[2] From *Composition of Foods*. Agriculture Handbook No. 8. (Washington, D.C.: U.S. Department of Agriculture, 1963).

cess during which the fish materials are treated with reagents that enrich the plant food contents.

As a consequence, the NPK grade of the end products may be regulated; for example, producing a high-quality 8-4-4 liquid fertilizer, containing about 6 lbs. of raw fish materials per gallon. Such a fertilizer has an immense potential in Midwest farming and gardening, straight or blended with seaweed as suggested in chapter 5.

Diluted with about 100 gallons of water, Carpole's Eco-Gro 8-4-4 is often used at the rate of one gallon per acre, sprayed on the foliage of crops.

FERTILIZER-GRADE FISH MEAL

The Carpole enterprise has a commendable flexibility that:

- if seasonal fish supplies are too great for producing fish meat, surplus whole fish may be made into liquid fish fertilizer
- if supplies are too great for making liquid fish, they can be converted to fertilizer-grade fish meal via a liming and composting process briefly described in previous chapters

Eventually, fish meal may be made in any case, due to the strategic value of this kind of fertilizer in stimulating soil bacteria in some kinds of farming and gardening. As research attention is given to this phase, it will surely be recognized as a primary energy-saver—using small amounts of fish and seaweed to cause bacteria to capture large amounts of nitrogen from the air.

FRESH WATER FISH FOR INLAND FARMERS

In America's Petroleum Age now closing, and previously, it was thought that fish fertilizers should come mainly from the sea for two reasons:

1. The marine supplies of fish are greater, and fish wastes accumulate at seacoast places.
2. Growing in oceans, marine fish are rich in minerals; richer, perhaps, than freshwater fish. Therefore, they may make better fertilizers.

Carpole's fish enterprise illustrates that neither reason is necessarily valid anymore. With carp on the scene, inland waters may yield enormous quantities of fish for use as fertilizers. Also, when carp and other freshwater fish fertilizers are blended with seaweed, as Carpole is doing under arrangements with Sea Born, Inc., the end product is "mineralized," just as if it came from the sea.

This area, too, needs research attention. Theoretically, a *complete*, fully efficient fertilizer can be made by blending 75 percent of carp-based fish with 25 percent or less of seaweed because:

* adequate nitrogen, phosphate, and potash are provided by the fish
* also, the fish provides a fair supply and assortment of other minerals
* the seaweed completes the mineral assortments, covering deficiencies that might be present in freshwater fish
* also, the seaweed provides cytokinins and auxins, which are needed in only "homeopathic" amounts, plus a useful chelating action in the composite product

Based on these premises, we believe inland farming and gardening may be adequately served with regional supplies of energy-saving fertilizers made from freshwater fish: carp, bullheads, sheepheads, other kinds, and dressing wastes from hundreds of catfish ponds. Blended with a bit of liquefied seaweed, fertilizers from these nearby freshwater fish may be just as good as those from coastal fisheries. Time and further research will tell.

ADJUSTING THE FLAVOR FACTOR IN HOPS

In 1955–1958, under arrangements with Ferro Corporation, trace mineral manufacturers, and a major U.S. beer company, we supplied fertilizers for many hop-growers of the Yakima Valley, Washington. The objectives in this case were to obtain high yields but also to sustain a gorgeous flavor in the hops. In doing this, we were dealing with *lupellin,* the aromatic flavor substance of hop berries, so highly prized by brewmasters.

In turn, *lupellin* is formed in hop berries under supervision of the plant's enzyme system, and, as we have noted, every enzyme has an earth mineral or sulphur as a key element in its chemical structure.

This we knew: too much regular nitrogen fertilizer would increase yield but depress *lupellin* formation in the hops. We had to achieve a nutritional balance so that both high yields and excellent flavor could be attained.

Our guides were sap tests of the hop vines during growth, measuring potash, magnesium, iron, and zinc. They served as tell-tale informers as to whether the plants' enzymes were fed, and enough hop fruit-making nutrients were flowing to the blossom cells. We ran over 1,000 sap tests in hop yards of the Yakima Valley, measuring nutrient levels in the vines.

Achieving the *lupellin* goals was rather easy, using these guides. We added fritted trace elements (FTE) and sulphate of potash-magnesia (Sul-po-mag) to the early spring fertilizers, and a chelated multielement spray (KE-MIN) a bit later, to sustain full mineral nutrition. Nitrogen and phosphate were left as before to maintain vine growth and yields.

In this program, the sap tests showed high iron and zinc *at lupellin-making time* in 90 percent of the vines tested, while other hop plants seldom showed enough for good flavor in the crop.

This is a good subject for farm research people to think about when sitting around drinking beer.

7.

PRACTICAL GUIDES, RECIPES, AND SUCCESS STORIES

In an earlier, better time for America's farmers and nurserymen, Clarence Prentiss ran a fine rhododendron and floral nursery in Seattle, Washington. Formerly a hearse driver in the undertaking industry, Clarence had a freedom from myths of the Department of Agriculture, and an inventive mind, lubricated now and then with Old Yellowstone whiskey. He proved to be a capable fellow investigator into how plants grow and use their foods.

Clarence said, "These plants out in the sun put flower buds on all their new shoots, but plants in the shade often fail to do so. Why?"

"Because," we said, "sunshine makes sugar via photosynthesis; raising the carbon/nitrogen ratio in the sap of the plants, sugar being COH, a carbon compound. And carbon is the sexy food. It makes plants want to bloom and have children."

"Hm-mm," murmured Clarence, sipping his Old Yellowstone, "then, why not fertilize with sugar?"

"Well, sugar on the ground wouldn't do it, and it could kill the plants. But sugar sprayed on the leaves might be okay. We will mix some up for you."

"Great," said Clarence, "we'll call it liquid sunshine."

At Tidewater Laboratory, we blended sucrose, phos-

101

phate, potash, and boron in a dilute solution and labeled it "Jet Power Liquid Sunshine." The next week, we made another pilgrimage to Prentiss Nursery, which was then on Airport Way, just opposite the old Boeing Company aircraft plant.

Assisted by Clarence Prentiss, we selected a bed of young rhododendron plants about 2 feet high, growing nicely in August. Down the center of the bed lengthwise we placed an 8-foot sheet of plywood and sprayed half the plants, sugar-feeding half of a bed 30 feet long.

Two months later, we counted buds in this bed of young rhodos and found we had increased the number of flower buds on the sprayed side by nearly 40 percent.

In that critical time when these plants were deciding whether to set flower buds or grow vegetative shoots, we had tipped the balance in favor of flower buds by raising the carbon versus nitrogen level in the sap. We were sunshine makers.

Next season Clarence Prentiss called and said, "Let's make a couple of moon shots. I have some ideas." Jet Power in hand, we went to the nursery and were shown a priceless Chinese rhododendron plant Clarence had brought to America seven years before and cared for lovingly. However, it had failed to set buds and bloom. This fine plant was about fifteen years old and 10 feet high.

We sprayed it three times in August-September with sugar, phosphate, potash and boron: COH along with nutrients to help the plant to use and translocate the sugar. And, Eureka! It grew buds and bloomed beautifully the following spring.

"Jeez!" said Clarence, "You are the sexiest fellows I have ever seen."

In that same year we sprayed a big potted pear tree, whose identical brother grew in a nearby pot-box 4 feet in diameter. Prentiss said, "Both will bloom and set a big crop of pears. Let's see if we can increase the yield of ripe pears on one of these trees."

In this case, we sprayed the tree when the little pears were ½ inch long, followed by three more spray feedings as they grew. And sure enough, we replaced sunshine and soil, enabling that tree to mature double the crop; twice as many pears as a young pear tree can possibly produce with normal nutrition. This silly fruit-loaded tree was exhibited by Prentiss Nursery at the Puyallup State Fair in October, 1955.

It was the year of Seattle's big freeze on November 11. In a lovely Indian summer, all kinds of plants had been growing. Dogwoods were blooming the second time, crazily thinking it was regular spring again. Then, on November 11, the temperature fell to 15°, and we had near-zero weather and snow for a week.

This proved to be our frost-resistance test on roses. At the Prentiss Nursery, we had sprayed fifty container-grown roses with "Jet Power," alongside several hundred unsprayed plants, to harden them for winter weather. And sure enough, none of these sprayed roses was killed by the November freeze, although all the rest died.

When nitrogen is the peer nutrient in the sap of plants, they think "grow some more." But when carbon dominates, and potassium paces the nitrogen, plants harden their tissues and get "fruit-minded." We gave that to the roses, and they survived the Big Freeze of 1955.

Finally, in this introductory narrative, let us consider boron, the "doorkeeper" in cell nutrition, and its twinship with calcium.

RHODODENDRONS HATE LIME . . . BUT . . .

In our work with "liquid sunshine" at Prentiss Nursery, we could not avoid learning more about boron, since it is the little "doorkeeper," helping sugar-making minerals to move effectively in plants. Along the way, we also learned a bit about lime and rhododendrons.

Ninety percent of all gardeners know, of course, that rhododendrons and azaleas *hate* lime. They are acid-loving plants. But, are they?

Already, we knew that calcium and boron are twin workers. They join together in forming cell walls, are critically essential *together* for sound plant growth.

When we were spraying rhododendrons with a boron-

103

carrying material . . . the liquid sunshine . . . Clarence Prentiss asked, "How much boron can these plants tolerate?"

Having to answer, "Nobody knows," we decided to find out.

First, we ran laboratory analyses of the calcium level in rhododendron leaves and, surprisingly, found it to be rather high, about the same as in apples or roses. That raised the question, "Why do rhododendrons hate lime if they take up all that calcium in their leaves?"

We reasoned: "Because they need more boron, to go along with the calcium in the lime." And we found this was true.

On a bed of young rhodos about 3 feet high, we sprayed various amounts of boron, including one group of plants at the "lethal" dosage of 20 lbs. of actual boron per acre. The plants loved it, responding with new growth of 4 inches or more.

In this reconnaissance work with sugar, boron, and multiminerals, we relearned that plant nutrition is indivisible. Each mineral nutrient is corelated to others, and still others, in twins, clusters, families, and whole galaxies of foods. Deficiencies of some minerals may only be excesses of others which are producing the deficiencies. Often plants and people seem to be just waiting for a few critical nutrients in order to grow and be healthy again.

As for the rhododendrons, we took some of these "acid-loving" plants into the laboratory in 1956 and grew them happily at pH 9.0, feeding limewater every week, *but* we had to give positive supplies of boron and an entire assortment of chelated minerals, especially iron. On such a *complete, balanced* diet these lime-haters became lime-lovers, the same as other plants, and grew normally.

Which brings us back to seaweed, the mineral-rich food from oceans for land plants. By 1957, we were making special fertilizers for nurserymen with seaweed meal as the wide-spectrum provider of forty minerals not commonly found in land people's fertilizers.

We learned that seaweed, too, can act as "liquid sunshine," to encourage fruiting, hardening, and frost resistance in plants.

GUIDES FOR USE OF SEAWEED AND FISH

Seaweed Meal

As a general guide, research work at Clemson University and elsewhere has shown 300 to 500 lbs. per acre of dry seaweed meal as a sound rate of use. However, this may be adjusted for different kinds of crops, soils, and climates. The main limitation is "saltiness" of the seaweed and some crops, such as potatoes and beets, thrive in mildly saline soils.

Converted to gardeners' and small farmers' terms, this rate of application equals about 1 lb. per 100 square feet, since 1/100 of an acre is about 435 square feet, and a pound for each such area amounts to 435 lbs. per acre. This quantity may be broadcast over the soil and mixed in by spading, roto-tilling or other cultivation. Or, the seaweed meal may be mixed with other fertilizer materials in a ratio that provides the desired application.

If seaweed meal costs 40 cents a pound in small bags or at local stores, this means a supply for an average-sized garden would cost about $3 to $4. Wholesale supplies for a farm, at about 25 cents a pound, might cost $80 to $100 per acre.

A reasonably dry seaweed compost may be applied at two to three times the above rate, since it contains more moisture and relatively less of seaweed substance, per se. Therefore, the general guide for people living near seashores, composting their own seaweed gathered from nearby beaches, is 800 to 1,200 lbs. per acre, or 2 to 3 lbs. per 100 square feet. Good luck, and we hope your crops, soil, climate, and seaweed are such that these rates are suitable. Local studies and experimentation may be useful.

Fertilizer Grade Fish Meal

When used as a stimulator for soil bacteria, as well as a fertilizer for direct use by plants, it is our experience that 200 to 250 lbs. per acre is a practical rate of application.

Growers of hops, fruit, and vegetable crops in Yakima Valley, Washington, successfully used such quantities of "live" compost as bacterial stimulants during 1955–1965, applying the material in or near the rows to get maximum effect at minimum cost.

We believe a similar guide is sound in the case of fertilizer-grade fish meal. Or, it may be used about the same as seaweed meal.

If the primary cost of fertilizer-grade fish meal, bagged and delivered in local areas, is around $250 per ton, it will have wholesale and retail costs for farmers and gardeners about the same as for seaweed meal.

Blended Fish and Seaweed Meal

It is our hope that a blend of fish meal and seaweed, along with a *nutritional* diluent, will be effective at lower cost than use of the straight materials. Examples of such *nutritional* diluents are: humates, lake-bed gypsum, rock phosphate, colloidal phosphate, feedlot compost, lime, composted sludge, and poultry waste compost.

If such blends prove to be efficient, the composite product might be used at rates from 200 to 500 lbs. per acre (½ to 1 lb. per 100 square feet of garden or nursery area) as a staple *energy-saving* fertilizer for many kinds of crops, working as a bacterial stimulant as well as a plant food.

The cost of such a product might then be reduced by 20 percent to 25 percent, compared with costs of straight fish and seaweed.

Research by business-minded people is needed in this sector. The net costs of these energy-saving fertilizers must be less than for petro-based products, or farmers and gardeners have no incentive to use them. These products must *substantially reduce needs for conventional fertilizers*.

Liquefied Seaweed

The traditional guide for use of liquefied seaweed is 1 gallon of the concentrate per acre, diluted with a convenient amount of water, such as 50 to 100 gallons. In the case of dry soluble seaweed powder, it is used at 10 to 16 ounces per

106

acre, reconstituted to liquid form and diluted in a similar convenient amount of water. Sometimes 12 ounces of powder is reconstituted into a concentrate (12 ounces per gallon of water) prior to dilution and spray application to the crop.

For foliar feeding, the leaves are wet thoroughly via spray equipment, and the surplus drips to the soil. None is wasted. The cool of the evening or early in the morning are preferred times of application. Midday in hot sun is much less efficient for foliar feeding of plants.

Timing for different kinds of crops is important. For example, soybeans should be fed seaweed at blooming time, and wheat when the young plants are making the second joint in their stems. Gardeners may spray at blooming time, and several times thereafter, often with excellent results.

Further specific guides for using liquid seaweed are given on product labels and in suppliers' brochures.

Costs of these materials vary from $5 to $20 per gallon of concentrate, and higher, depending on the types of products and claims of manufacturers. However, farmers may expect to pay about $6 to $10 per acre per application, and gardeners may get enough liquid seaweed for a season for about $2.50.

Lacto-base Cultured Seaweed

Almost by accident, midwest seaweed companies and farmers have learned that *lactobacillus acidophilus* grown in a seaweed culture make a magically effective liquid fertilizer. This is the same group of bacteria used to make yogurt, so useful in animal and human digestion. However, in the case of lacto-base seaweed fertilizers, the bacteria are not in live condition; they are dead.

The lacto-base material is rather special, since remarkably small amounts per acre may cause substantial increases in crop yields. Currently, the suggested rate of application is about 8 ounces (½ pint) per acre, diluted with water, sprayed on such crops as soybeans, cotton, wheat, corn, and rice. Side-by-side comparisons of yields from treated and untreated areas have shown increases of 10 percent to 40 percent; but sometimes zero for unknown reasons.

EVERY YEAR A VINTAGE YEAR

Since the times of the old monasteries, vintners have awaited anxiously with one eye on the weather and the other on moon signs to see if the wine was good and . . . hope of hopes . . . if it was a *vintage year*.

Well, what is a vintage year and why does it happen?

Let us say it this way: In a vintage year, the ancient chromosomes and gene strips of the grape variety have a climax event. They put into the grapes *all* of the genetic capabilities of the variety, evolved in misty eons of the past. Precious, gorgeous, superb. A vintage year!

But why not every year? It is mainly a matter of skillfully fertilizing the grape vines. In principle, these are the guides:

1. Grow Strong Vines. Sustain this part so the leaf surface is adequate to make enough sugar to raise the soluble solids in the grape juice above 20 percent. Shoots should grow to 55 or 60 inches on some varieties to attain this goal. This phase can be governed within 10 percent by fertilization.

2. Feed the Seeds. By mid-August, the grape seeds are forming, yelling, "Hey, send us magnesium, phosphate, and more minerals!" So, there must be enough mobile magnesium in the plants for seed making, without robbing the leaves of chlorophyll needed to keep on making sugar (chlorophyll is 4 percent magnesium). The plant must "keep on truckin' " in the sugar department, no matter what happens in the seed-making department.

3. Feed the Enzymes. They supervise cell building and growth, but also formation of aroma and flavor factors. Feed them *all* of the known and unknown mineral elements so that they can perform these delicate end-result functions. How? With regular chelated trace minerals, *plus* seaweed. These *are* the moon factors.

As for the weather, the potash-magnesium-mineral foods, *are* weather. Skillfully handled, they can act in lieu of sunshine.

Finally, sometimes the vintage year was there, but transient molds came on the grapes and spoiled it. Seaweed *may* prevent this. Remember Dr. Senn's work on peaches. He controlled mold at harvest time by seaweed sprays in June. It is possible that this technology can also be used advantageously on grapes, controlling molds while simultaneously feeding the enzymes.

This, too, is a good subject for wine buffs to think about as they roll the new wine around in the glass, sniffing and savoring it, seeing if it may have those errant values that make a vintage year.

The farm price of the lacto-base liquefied seaweed is from $90 to $110 per gallon, or about $6 per acre when the product is used at the rate of 8 ounces.

If and when lacto-base cultured seaweed is shown by research evidence to cause growth increases when used so sparingly, the effect will have to be attributed to hormonal and/or enzyme factors, since the amounts of conventional nutrients in 8 ounces of lacto-base seaweed are negligible.

Guides for applying and using the lacto-base materials are given on product labels and in suppliers' brochures.

RECIPES AND HOW-TO-DO-ITS

Wine-makers, cooks, lovers, and fertilizer people have traits in common: as artists, they help Nature, then unashamedly take credit for Nature's work. As liars, they believe their own stories; therefore, are the world's most honest people. With that warning, we shall share recipes with you.

Liquid Sunshine: Bud Builder and Plant Hardener

The sugar-based "liquid sunshine" is a bit difficult for lay people to make and use in its fully technical formulation; however, a simplified version can be made by any gardener or nursery person. Based on seaweed, it provides nutrients that help plants to harden their tissues, set flower buds, resist frost damage, increase yields, and repel insects. The recipe is as follows for 1 quart quantities. Farmers and nurserymen can multiply the amounts by 10, 100, or 1,000 depending on the sizes of their operations.

1 pint	Corn syrup, such as Karo (contains 500 grams sugar)
2 ounces	Liquid seaweed extract

110

1 pint	Water
1 teaspoon	Boric acid
1 quart (approx.)	Total

Mix well.

Directions: Dilute 4 tablespoons of "liquid sunshine" per gallon of water. Spray on foliage of plants until wet. For tomatoes, spray 3 times at weekly intervals, beginning in July or as tomatoes begin to form on plants. For rhododendrons, azaleas, and other broad-leaf evergreens, spray three times at weekly intervals, beginning in late August. To harden roses and other plants and prevent frost damage, spray three times, at weekly intervals, commencing one month before expected first killing frost.

Good luck. You are a pioneer, and on your own! We don't carry malpractice insurance.

"Dry" Liquid Sunshine: Bud Builder and Plant Hardener

Based on similar principles, this special dry fertilizer also has capabilities for helping plants to harden their tissues, build buds, resist frost, increase yields, and repel pests. In this formulation seaweed acts as a sugar-maker and supplier, along with the other ingredients. The carrageenen in seaweed is a carbohydrate . . . a sugar of the sea . . . acting in place of corn syrup in the liquid formulation. And, both seaweed and FTE are boron-suppliers. Here is the recipe:

10 lbs.	Dry compost and/or peat moss, or other organic base material
20 lbs.	Dry seaweed meal
5 lbs.	Bone meal or single superphosphate
10 lbs.	Sulphate of potash-magnesia (Sul-po-mag—see Appendix), or, 5 lbs. sulphate of potash

111

1 lb.	Fritted trace elements (FTE—see Appendix)
4 lbs.	Agricultural gypsum or lime

50 lbs.	Total

Put ingredients in a pile and mist thoroughly. If you use 5 lbs. of sulphate of potash instead of 10 lbs. of Sul-po-mag, just add another 5 lbs. of compost to make up the difference. If you cannot get FTE, leave it out and add compost. Be practical, use your imagination. Don't get frustrated.

Apply the bud builder broadcast over the root feeding area in late summer or early fall. For shrubs, roses, and small trees, apply one *small* handful per foot of height of plant. For fruit trees, apply 1 to 5 lbs. per tree, depending on size. For berries and gardens, apply 1 lb. per 100 square feet of area.

Ecological Garden Fertilizer[1]— Nickname: Deluxe Eco-Grow

To make this all-purpose garden fertilizer, we suggest using any good organic-based fertilizer as a primary ingredient, since many readers would otherwise have difficulty finding and buying fertilizer materials in their own neighborhoods. Then, seaweed and other special ingredients may be added. Here is the recipe: *

10 lbs.	Any good organic-based garden fertilizer, such as 5–10–10 or 6–10–4. Or, a rose fertilizer may be used
5 lbs.	Ureaform or "organiform"
5 lbs.	Bone meal, or superphosphate if bone meal is unavailable
4 lbs.	Sulphate of potash, *and*

[1] *From Ecological Gardening for Home Foods* by Lee Fryer and Dick Simmons. New York: Mason/Charter Publishers, 1975.
* See Appendix for descriptions of some of these ingredients.

112

1 lb.	Epsom salts, *or*
(5 lbs.)	Sulphate of potash magnesia * (Sul-po-mag)
10 lbs.	Seaweed meal
15 lbs.	Compost or dry manure

50 lbs.	Total

Pour the ingredients in a pile and mix thoroughly. If needed to increase bulk or reduce dust, add 5 lbs. of damp peat moss or garden mulch.

Use at rate of 4 lbs. per 100 square feet of garden area, tilled into soil. Also, may be sowed under rows when planting seeds or transplants. For shrubs and bushes, use one small handful per foot of height of plant; for trees, use 1 to 5 lbs. per tree depending on size.

Full Organic Garden Fertilizer[2]: Nickname: Eco-Organic

To avoid problems finding various materials, this fertilizer, too, is made with any locally available organic fertilizer as a primary ingredient. Here is a recipe:

15 lbs.	Any good 100 percent organic garden fertilizer available in your area; such as a 5–8–5 or 5–10–10
10 lbs.	Ground rock phosphate or bone meal; or both
15 lbs.	Compost or dry manure
10 lbs.	Seaweed meal

50 lbs.	Total

Mix thoroughly and use as described above for Eco-Grow.

* Sulphate of potash-magnesia contains both sulphate of potash and Epsom salts, therefore may be used in place of these two ingredients.
[2] See previous citation

**Seaweed and
Fish Blend:
Dry
Ingredients**

To make this deluxe fertilizer, we specify *more* fish and *less* seaweed, since: (a) fish provides the essential nitrogen, phosphate, and potash, plus other minerals, (b) seaweed supplies hormones and chelates that are effective in lesser amounts, and (c) seaweed's trace minerals, beyond those provided by fish, are required in only tiny quantities, if at all. Therefore, we suggest a ratio of about ⅔ fish and ⅓ seaweed in a straight blend of these materials.

However, since seaweed and fish are concentrated materials and quite costly, we suggest that other locally available materials may be blended with them. The following recipe may be used:

10 lbs.	Dry seaweed meal
20 lbs.	Feed-grade fish meal
20 lbs.	Any lower cost bulk material, such as agricultural lime (dolomite), gypsum, humate, rock phosphate, compost, or dry manure

50 lbs.   Total

Pour these ingredients in a pile and mix thoroughly with a shovel or other implement.

If fertilizer-grade fish meal is available, it may already contain a suitable diluent. In that case, simply mix 75 percent fertilizer-grade fish meal and 25 percent seaweed meal.

Use at a rate of ½ to 1 lb. per 100 square feet of soil area (200 to 500 lbs. per acre) to stimulate and nourish soil bacteria, as well as the plants and crops.

**Liquid
Seaweed
and
Fish Blend**

To make this deluxe liquid fertilizer, the following guides should be observed:

114

1. Don't mix alkaline fish (pH of 7.0 or higher) with alkaline seaweed. They are incompatible.

2. SM-3 liquid seaweed has an acid reaction (pH below 7.0) and may be mixed with any kind of liquid fish, including fish solubles (fish "emulsion"). Fish solubles are alkaline.

3. Carpole's liquid fish is of acid reaction, and may be blended with all kinds of liquid seaweed, including Maxicrop, Sea Born, Sea Crop, SM-3 and Sea Spraa.

With the above guides, we suggest the following recipe:

75 percent liquid fish
25 percent liquid seaweed
-----
100 percent Total

Mix thoroughly with a paddle, beater, pump, or outboard motor, depending on quantity.

This deluxe liquid fertilizer may be used at the rate of 1 gallon per acre, mixed with 100 gallons of water, or with the amount of water convenient for spray equipment. Spray on foliage of plants until they are thoroughly wet.

Add WEX, TWEEN, Triton, or any other good wetting agent (surfactant), if available, following instructions on the label. It will promote effective action when the seaweed/fish blend is sprayed on the plants.

**Seaweed
Compost**

Seaweed contains insufficient nitrogen for making a good compost, so nitrogenous materials should be added. Remember, a good carbon/nitrogen ratio for composting is 10 to 1; and, as is, seaweed contains less than 1 percent nitrogen. The following guides may be used for making seaweed compost:

600 lbs.	Wet seaweed from beach or shore
200 lbs.	Manure, compost, fish, feathers, slaughter waste, sludge, or other nitrogenous wastes
100 lbs.	Agricultural lime and/or gypsum

115

100 lbs.	Rock phosphate or other phosphate
1,000 lbs.	Total

Mix thoroughly, handle and turn as when making any other kind of compost. Loamy soil may be added to the pile, if desired. Adding rock phosphate or other phosphate material is desirable, since seaweed and the nitrogenous wastes are low in this essential food, and compost bacteria need phosphate while doing their work.

The finished compost may be screened and coarser chunks returned to the next compost piles.

Foliage Brightener: A Chlorophyll-Maker

When plants get tired in summertime, they may lose green color (chlorophyll) in their leaves; and photosynthesis (sugarmaking) may therefore decline sharply. Supplies of iron which flowed abundantly in April have dwindled. Magnesium is moving from chlorophyll cells to fruiting sites, to help in making seeds, leaving the leaves wan and pale between their veins (chlorosis), and nitrogen, the always-needed nutrient, is in low supply.

To restore color and vitality in such plants without including soft growth to be damaged by frost, a seaweed-based liquid fertilizer may be used. The recipe is as follows:

4 oz.	Iron chelate: ask for Geigy's, KE-MIN, or other good product at your garden store
1 lb.	Epsom salts (magnesium sulphate)
8 oz.	Liquid seaweed concentrate
4 oz.	Potassium nitrate; call a chemical supply company

To mix, fill a gallon jug half full of warm water and add all the ingredients. Stopper and shake well until all are dissolved. Then fill the jug with more water. This makes 1 gallon of liquid concentrate.

116

Dilute 1 cup of this concentrate with 1 gallon of water to make the spray material. WEX, Triton, or other wetting agent may be added to assist in application to plants. Spray them once or twice. They will reward you by greening nicely and getting happy again.

In this "leaf brightener," you are using the cytokinin technology of Gerald Blunden and others, who have shown that liquid seaweed improves efficiency of photosynthesis in plants. Also, you are providing iron and magnesium, the chlorophyll makers, along with potash and boron—and a wee bit of nitrogen to sustain the plant in its autumn seed-building operations.

For more information about the materials and ingredients for these recipes see the Appendix A.

SUCCESS STORIES

In the U.S. farm research system, the state and area experiment stations tend to devote their time and resources to products and technologies that are blessed by national farm research leaders, to the exclusion of other products and technologies. Also, financing of specific research projects is provided, in part, by private grants and contributions, including those of fertilizer, petroleum, farm chemical and other interested companies; whereupon these popular externally supported projects tend to preempt the time of capable research personnel, as well as the facilities of the experiment stations.

Since seaweed research has not been blessed by national farm leaders, nor supported adequately with grants and contributions, very little of it has been conducted in the U.S. The pioneer work in this field at Clemson University and at the University of Maryland (described in chapter 9) was not supported by USDA and it has had only minor financial support by private companies, mainly Sea Born Corporation and the Maxicrop Company.

However, many capable farmers have joined with responsible seaweed companies in conducting practical tests of seaweed products in various crops under field conditions, in some instances over a period of years. Glen Graber of Hartville, Ohio, is an example of this, developing guides and success models observed by hundreds of visiting farmers and agriculturalists, even some from England and other foreign countries. His operations were described in chapter 1.

To provide readers with further insights into practical uses of seaweed in growing farm crops, we will now give additional examples under the heading of "Success Stories." They have not been validated or replicated, or made to conform to approved research designs. However, we believe they are reasonably accurate and useful as guides for other growers.

YIELD INCREASES IN SOYBEANS
WITH LIQUID SEAWEED

In 1975, at least five hundred Midwest farmers sprayed their crops with liquid seaweed, because experience in their neighborhoods showed the profit-making value of doing so.

Leonard Tjaden, Route 4, Charles City, Iowa, is a good example. In a 24-acre field of soybeans, he sprayed half the field at blossom time with Sea Born liquid seaweed, reconstituted from soluble powder, at the rate of about 11 ounces per acre. The cost for seaweed was approximately $4.00. Here are the results:

	SOYBEAN YIELD PER ACRE
12 acres sprayed with liquid seaweed	54.0 bushels
12 acres unsprayed	46.9 bushels
Increase	7.1 bushels
Percent increase	15%
Value per bushel	$3.00
Value of increase per acre	$21.30
Cost for seaweed and spray application	$6.50
Net profit per acre	$14.80

Eugene Holzer, also of the Charles City area, reported a similar success using soluble seaweed powder on soybeans. In a field of 108 acres, he sprayed half the field, 54 acres, at 11 ounces per acre diluted in water. The untreated portion produced an average of 24 bushels per acre; the sprayed part 28 bushels, an increase of 4 bushels or 16 percent. In his case, the net profit from using liquid seaweed was about 6 dollars per acre.

Commenting on these success events and others like them in wheat, soybeans, corn, rice, and cotton, Paul Wolfgang, general manager of Sea Born, Inc., Charles City, Iowa, said, "When these farmers started using liquid seaweed, it was to get higher yields at low cost and thereby add to their net incomes. That is still a major factor, but now the energy problem is entering the picture, too. Nitrogen costs are rising and future supplies are uncertain. Farmers know this. They are interested in energy-saving as well as cost-saving, and seaweed and fish are the best energy-savers in sight."

RESURRECTING A FLORIDA ORANGE GROVE WITH SEAWEED

In 1973, Bob Plimpton of North Palm Beach, Florida, said, "To heck with the public relations business. I'm going to be a citrus farmer."

By stretching savings, credit, goodwill and all else, he was able to buy Palm Beach Ranch, a 1600-acre citrus property near Palm Beach, of which 800 acres was in oranges and 800 acres in lemons. It helped in making the deal that the trees were malnourished, bug-infested and in generally poor condition. They had citrus decline, a widespread problem in the Florida citrus industry.

During 1974–1975, Bob Plimpton eliminated all conventional chemical fertilizers and toxic pesticides. "I really did it too fast," he says. "It would have been better money-wise if we had made a slower transition. But we took drastic steps, and I am glad we did."

Plimpton found a useful low-cost bulk organic material: sugar-mill waste. Composting this, he applied 10 cubic yards per acre (about 5 tons) in two applications. Also, knowing about seaweed, he went to England and studied the SM-3 seaweed products made by Chase Organics, Ltd., at Middlesex; and he became America's leading importer-distributor of the SM-3 seaweed materials.

Commencing in 1974, the 1,600 acres of citrus were sprayed with SM-3 liquefied seaweed in early spring at the rate of 1 imperial gallon per acre. This treatment was supplemented by a postbloom liquid feeding of chelated iron, copper, manganese, zinc, and boron; and soil feeding of sulphate of potash-magnesia (Sul-po-mag) along with the compost.

To neighboring growers the results seem to be miraculous, but they are really what might be expected.

"Now I have a fairly clean and healthy citrus grove," says Bob Plimpton, "and it was the sorriest mess you ever saw. Snow scale, white flies, and gumosis were rampant and many of the trees were dying. The only pesticide I have to use now is Banex, a nonpoisonous spray for insects. The SM-3 seaweed and good mineral nutrition have reduced the pest populations by 80 percent and Banex does the rest."

Bob Plimpton is controlling tree decline, the scourge of Florida citrus growers. As he says, the tree decline is only the end result of poor farming practices, relying too much on petrochemical fertilizers and toxic pesticides. The seaweed program provides the tools for shifting to a better, safer system of farming.

The red cherry on top of Bob Plimpton's "success sundae" is a deluxe market. By cutting out the toxic pesticides and "chemical" fertilizers, he has qualified the orange crop for the organic market, and Americans are famished for safe good-tasting foods. Using a special quick-freeze process, Plimpton packages 100 percent pure organic juice from his rejuvenated trees, and sells this through Shiloh Farms to 1,150 health-food stores all over the United States.

In only four years, a capable new farmer has built a success event raising superoranges in Florida, using seaweed as the key to unlock closed doors.

You will find Bob Plimpton's new seaweed distributing

120

company listed in the Appendix as Atlantic & Pacific Research, Inc., North Palm Beach, Florida.

A MILLION TOMATOES PER ACRE WITH SEAWEED

First, let us set our sights and measuring sticks: a regular home tomato plant will often grow 3 feet high and produce a dozen ripe tomatoes, using about 10 square feet of soil area. At this rate, you could put about 4,500 tomato plants on an acre of land and get a crop of 54,000 tomatoes. If they weighed 6 ounces each—fairly big ones—this would amount to about 20,000 pounds, or a yield of 10 tons per acre.

Good commercial tomato growers often get more than twice that much, from 25 to 30 tons per acre of smaller sized fruit, feeling happy to produce 200,000 ripe tomatoes per acre in a growing season.

Then, how about this million tomatoes grown with seaweed?

It happened this way. As a retired steel company executive, Jim Wagner was irked with urban life, so he moved to Capitola, California, and began to help Chicano farm workers to raise their own vegetable crops. He studied tomatoes and cucumbers with the fresh mind of a highly trained nonfarmer, a steelmaker.

"Why raise this valuable crop outdoors at all?" he asked. "Too many hazards, bugs, and variables. Let's just put it indoors."

With an Economic Development Administration grant, Jim Wagner helped the Chicano co-op to build new-style hydroculture units; not greenhouses or hydroponic facilities, but simpler, lower-cost structures in which air, temperature, light, water, and fertilizers could be controlled while the tomatoes grew in the ground.

He came to us for fertilizer guides, and we showed him seaweed. Nighttimes he studied *ABC's of Agrobiology* by O. W. Willcox—the bible on how to grow maximum crops—and other horticultural thrillers. In step-by-step fashion, Jim Wagner learned how to use low-cost organic wastes and com-

121

bine them with strategic crop feeders like seaweed, in growing superb yields of vegetable crops, especially tomatoes.

With spent mushroom compost from California's mushroom industry, he built a superrich soil, and fertilized it with seaweed meal, sulphate of potash-magnesia (Sul-po-mag), lime, and phosphate. Then, under a plastic roof with well-controlled air, temperature, and water, he planted tomatoes, and along each row was a low-pressure dribble feeder to provide liquid nutrition in the irrigation water.

In this kind of "hydroculture" operation in 1974, Jim Wagner grew 160 tons of tomatoes per acre, about one million ripe tomatoes of commercial size. Filled with vitamins and minerals, these tomatoes had super-flavor and long shelf life. Supermarkets and vegetable brokers paid a 20 percent premium for them.

Where is Jim Wagner now? He is general manager of Earth & Sea Products, Inc., at Watsonville, California. Recaptured by our business system, he is moving successfully into the seaweed and fish-based end of the fertilizer business.

Speaking of fertilizer people, let us close this chapter with the epitaph that used to hang on our wall:

Old fertilizer men never die,

They just smell that way.

HOW TO GROW YOUR NAME ON AN APPLE

An apple sitting on your kitchen table gives off a pleasant scent, and that's the key to this story. With apple scent, you can grow your name on another apple.

We learned this in 1955–1960, when we supplied complete mineral fertilizers for big apple-growers of Yakima and Wenatchee, Washington, and shared knowledge with Verle Woods of Crop King Company.

Apple growers of that region store their fruit in big refrigerated warehouses in which 100,000 or more boxes may be kept. Imagine the scent from all those apples! However, the scent-laden air is drawn off in ventilation systems through charcoal-filled cannisters because apple growers learned long ago that the fruit would keep better if this were done. One of Crop King's services was cleaning the charcoal cannisters, and reactivating the charcoal so it could catch more apple scent.

Said Verle Woods, "That apple scent must be interesting stuff. Let's fractionate it and see what we've got." And he did that.

Distilled at various temperatures, the scent oil released over twenty different chemical fractions. Most of these had no apple scent at all, but possessed remarkable abilities to do things to apples. For example, there was a "rotting" fraction that would rot firm, fresh apples in a matter of minutes by providing enzymes that destroyed pectin in the cell walls.

Other fractions killed fungi, repelled insects, and controlled nematodes in experimental applications. One component of apple scent turned starch into sugar in whole apples.

However, the color fraction was the most dramatic of all. With this part of the scent, we could grow our names on apples in just a day or two. All we did was make little stencils saying "Dick," "Lee," or "Mary" and paint the words on apples hanging on trees, whereupon the apples would grow our names, using potent enzymes from this fraction of the scent.

Yesterday, this was a novelty; today, it is almost forgotten. Tomorrow? Who knows? As Svengali said, "There are powers and forces in this world we've scarcely dreamed of."

8.

FARMING AND GARDENING IN 1985 WITH SEAWEED, FISH, AND ORGANIC WASTES

A few years ago (up to 1950) America had 6 million farms and ranches scattered everywhere from coast to coast. Over half of them were livestock or general farms with beef cattle, milk cows, pigs, or poultry. The animals ate pasture grass in 3 million different fields and places, scattering their manure decently over the earth, recycling the wastes, polluting nothing.

In nonpasture season, farmers fed the animals and birds in more than 3 million different places, and had small manure piles which they hauled out to 4 or 5 million fields as fertilizer—local energy.

That was *ecological* agriculture, conserving resources and energy due to having many small decentralized production units at several million different places.

Along about 1950, big changes were made: Hy-Grade Meat Company and many others built big beef feedlots to squirt DES (di-stilbestrol) into the animals and feed thousands of them in central places, instead of locally near the pasture lands.

Result: Big manure piles.

Ralston-Purina and many other companies developed big poultry colonies of half-a-million birds apiece, to eat their *high energy* feeds, instead of having the chickens out on a million different farms in small flocks.

Result: Big poultry manure piles and massive accumulations of poultry-dressing wastes.

Carnation Company and other companies built large two thousand cow dairies in Wisconsin, Florida, California, and other big market zones, and shipped milk in tank cars like petroleum for 500 miles or more. The milk you drank in Mobile, Alabama, was produced in Illinois, rather than at a nearby dairy. Same for Philadelphia.

Result: Mountains of cow manure at the big dairies and no place to put it; so much it would drown nearby fields in filth rather than properly fertilizing the crops.

In this changeover period, the Schaake Meat Packing Company of Ellensburg, Washington, set up a beef feedlot alongside the Yakima River to fatten thousands of beef animals grown on nearby pasture lands.

Result: A big manure pile in that lovely valley, and also a blood disposal problem. Butchering his own beef cattle, Schaake first poured the blood into the Yakima River but this upset the ecology and polluted the clear waters. An order was issued causing Schaake to cease and desist putting feedlot and meat-packing wastes into the Yakima River.

A responsible, civic-minded man, Schaake came to us and said, "Can you compost these wastes and sell them as fertilizers?"

With a big fertilizer-mixing plant at Ellensburg, plus others in the Northwest region, we said, "Yes, if you will pay the development costs, we will compost your feedlot manure, blood, and other wastes and move them into the fertilizer market."

We did that during 1955–1960, and learned a great deal about the possibilities and problems in recycling and using massive organic wastes.

MAKING COMPOST AT SCHAAKE'S FEEDLOT

Our first move at Schaake's was to hire Bill Turney, Yakima's Public Health Officer, since he was one of the region's finest experts in organic chemistry, sanitation, and waste disposal. Having served as Sanitation Officer in the Canal Zone and other places, and studied Berkeley's garbage composting project, Bill Turney knew how to solve Schaake's waste problems.

First, we executed a business agreement with Schaake to buy all of his finished compost, dried and screened to ⅛-inch mesh, for 12 dollars per ton FOB Ellensburg. This gave him a sound business basis for a composting enterprise, and gave us a low-cost supply of fertilizer materials.

With large-scale equipment, Bill Turney piled the feedlot manure and wastes in windrows, 100 feet long, 12 feet wide and 7 feet high, each with a lengthways furrow on top. From a tank truck, he poured slaughterhouse blood into the furrow, anointing each pile with this wild nitrogenous substance. Remember, blood is 16 percent nitrogen, dry basis, an excellent fertilizer but a reeking nuisance if neglected.

Within a few hours, the wastes began to "cook." Fed by the nitrogen and other foods, trillions of bacteria consumed lignins and other fibrous materials giving off heat and changing the mass to black humus. But if the temperature rose too high, pockets of "fire-fang" developed in the compost piles, so Bill Turney kept long thermometers in them to signal rising heat. At 160°, cool water was added and the piles were turned to assure thorough decomposition of all the materials.

After three turnings and three or four weeks' time, the feedlot wastes anointed with blood were converted to a black, loamy "live earth" containing about 30 percent moisture. Screened to a ⅛-inch particle size, it was a nurseryman's dream for potting and fertilizing plants. Also, it was a popular gardener's item, sold in 50- and 25-lb. bags.

However, 30 percent moisture is too high for a bulk organic "base" to use in making lawn and garden fertilizers. Mixed with other ingredients, the damp compost would cause chemical reactions and heating in the products. Therefore, we set up a big drying facility to reduce moisture in the compost to about 12 percent, opening up a still larger market for the Schaake wastes, as an organic ingredient in making many kinds of lawn and garden fertilizers. Even our competitors were interested in buying this superb compost.

127

By taking these various steps, we created the business base for recycling Schaake's feedlot and slaughter wastes. Instead of a limited market among local farmers and gardeners, the organic materials could reach farms, nurseries and gardens in a 500-mile radius, and even in Alaska and Western Canada. The Schaake compost was sold in the following forms and markets:

- in bulk to farmers
- in bulk to nurserymen and landscapers
- in bulk to commercial turf people for use on golf courses, athletic fields, cemeteries, and public areas
- in utility-type bags to farmers, nurserymen, landscapers, and turf people
- bagged for home gardeners
- as an organic base material in many kinds of mixed fertilizers

In 1955–1960, the aggregate of these uses and markets was sufficient to consume the organic wastes of Schaake's big beef livestock feedlot. For a brief time, a successful model of processing and recycling a large volume of farm wastes was demonstrated; then this model was destroyed by competition of Ortho and other oil company fertilizers. While it lasted, these were the factors accounting for success:

1. A civic-minded feedlot operator, stimulated by a court order to cease pollution of water and land.

2. A large-volume fertilizer company,* capable of marketing the composted wastes into a wide area.

3. A supercompetent technologist and site manager, Bill Turney.

4. A diversified product and marketing program, designed to put the compost into many forms for many markets.

5. A deliberate process of *upgrading* the value of the product so it could sustain freight charges and be shipped profitably for long distances, even into Oregon, Idaho, Western Canada, and Alaska.

However, that was *then*. How about *now*, 20 years later?

* The Chas. H. Lilly Company, Seattle, Washington

WILD BLOOD

In the days of Upton Sinclair's *The Jungle*, beef cattle and hogs were fattened on farms and ranches, shipped to Chicago and other meat processing cities and butchered there in the stockyards. It made a river of blood, wild stuff, but this by-product was quickly made into paints, adhesives, feeds, fertilizers, and a dozen other useful end products. Roughly speaking, sound ecological principles were expressed in the country's meat-producing industry.

Then along came the big feedlot operators, saying, "We'll fatten the beef right out here on the plains of Colorado, Kansas, and Texas, near the grain and feed supplies, and save everybody a lot of money, while we make a couple of bucks ourselves."

The rest is history. They did it, and thereby destroyed the packinghouse industries of Chicago, Omaha, Sioux City, and other meat industry centers. However, another thing was wrong, too: there was no place out at the feedlots to put the rivers of blood formerly made into useful products, or the mountains of manure from the cattle who formerly left it on a million farms and ranches in the grain and livestock country . . . scattered and safe, polluting nothing.

It is the blood that is explosively wild. Ninety-five percent protein, corpuscular, and full of life's foods, it rots stinkingly *right now!* Swarms of blow flies lay eggs in it and the maggots squirm disgustingly. Run it into a stream and it kills the fish. Pour it on the land and it destroys all life.

One time a Western feedlot operator who butchered his own cattle tried hauling the blood away in tank trucks but one of them overturned in a town of about 6,000 people. Disaster! Blood all over the streets. That town was hardly worth living in anymore. Everyone remembers when it stank like a battlefield from its big blood bath.

Wild blood. Every butchered cow yields about 4 gallons of it, every hog a gallon, and poultry put out about 13 gallons for every 1,000 chickens we buy in food markets or eat in the old colonel's fried chicken palaces. Add the calves, sheep, and turkeys and it makes about 300,000,000 gallons a year to use or waste . . . and wasting is no longer possible without seriously damaging land and water.

That is the nation's blood problem, and it is time to get on top of it. One way is to "go vegetarian," thereby curtailing the slaughter of animals and birds, but that way seems unlikely. The other is to use all of the blood, tame it; and we know how to do that, since blood is 16 percent nitrogen, dry basis. It fertilized the Tree of Liberty, and it can also grow beautiful food crops.

If America must "kick the petroleum habit" and shift increasingly to "organic" fertilizers and pest controls, can this be done efficiently, without cutting crop yields and raising food costs? Can America be fed adequately from farms and gardens fertilized with seaweed, fish, and the wastes of farms, feedlots, and cities?

We will consider these questions in this chapter, and try to foresee how crops may be grown in 1985, less than ten years from now. One thing is certain: they *will not* be fertilized in the fashion of 1975, because the country is running out of gas and oil. The trend is toward Schaake-type compost units, rather than away from such waste conversion enterprises.

NEW VALUES FOR ENERGY-SAVING FERTILIZERS

Albert Einstein and his fellow physicists gave us curved space and new time machines to measure it. In a similar fashion, agronomists will have to develop modernized waste-based farming to supersede petro-based farming, and a new value system for the essential fertilizers. The old guides and state fertilizer laws are becoming as obsolete as the dodo.

For perspective on this phase of the problem, we may refer again to costs and values in 1955, when we were making and selling Schaake compost. At that time sulphate of ammonia, 21 percent nitrogen, cost $63 per ton or $3 per unit (percent) of nitrogen; and this was generally recognized by fertilizer makers to be the value of nitrogen in any fertilizer material. In a similar way, values were set for phosphate (P_2O_5) at about $1.50 per unit (percent in the product), and 70 cents for potash (K_2O). That is the way fertilizer-business people evaluated any and all fertilizer materials.

Using these guides, the value of Schaake compost, rated at 2 percent nitrogen, 1 percent phosphate and 1 percent potash, was only $8.20 per ton. However, hop farmers, fruit growers, nurserymen, and golf greenkeepers were willing to pay $50 per ton, since they found nonrated values in this compost product. It stimulated soil bacteria, increased yields, and improved the quality of their crops, so they were willing to pay the premium price.

Currently (1976), fertilizer-grade urea with 46 percent nitrogen costs $145 per ton wholesale, and ammonium nitrate with 33½ percent nitrogen is $110, indicating that nitrogen is still worth about $3 per unit (percent in the product).

However, this nitrogen value, based in petrochemical supplies and costs, is quite unstable and subject to rapid fluctuations upward. For example, during the mild "energy crisis" of 1973–1974, urea quickly rose to $240 per ton or nearly $6 per unit of nitrogen. For a while, many farmers and fertilizer companies were unable to buy urea and other forms of concentrated nitrogen except at exorbitant prices in "black markets."

That was only a gentle prelude for what will happen when natural gas gets really scarce and rises to $2 per thousand cubic feet.

An immediate effect is to enhance the values of compost, manures, crop wastes, sludge, seaweed, and fish. The forces of Energy Age economies are immediately felt, attracting these energy-laden materials into our production system, superseding and outmoding the economics of the Petroleum Age.

PRESTO! THE VALUES ARE THERE

Our working figure for nitrogen value in this illustration is $6 per unit. Since one unit is 1 percent of a ton, in fertilizer people's language, it is 1 percent × 2,000 lbs. = 20 lbs. of *actual* elemental nitrogen. Such nitrogen therefore has a value of 30 cents per pound (6 dollars divided by 20 lbs.).

When fertilizer costs rise to this level, it is true that the NPK values of compost, sludge, and fish rise, too. However, the increases are not nearly enough to attract these materials into large-volume fertilizer markets. Even if we add 50 percent because these materials are "organic," the values are only as follows:

Good feedlot compost (2-1-1) is worth only about $22 per ton—less than enough to make and bag the product to say nothing of hauling it 100 miles.

Sewage sludge (2-1-0) is worth only $20 per ton—too little to pay for processing and hauling it into farm areas from city sanitation plants.

Fertilizer-grade fish meal (10-4-1) is worth only $100 per ton—hardly enough to pay the costs of drying, grinding, and hauling it from fisheries to farm areas.

Seaweed meal (0-0-3) is worth only $10 or $15 per ton based on its potash and mineral contents—too little even to pay for gathering it.

Yet, these are the strategic kinds of organic materials that must be recovered and cycled into farming and gardening if America is to cope with an impending *real* energy crisis. How can it be done?

This is the way: Count in the bacterial and other nonrecognized factors, and presto, the values are there! Here are some examples:

Bacterial Action

If a compost material used at 500 lbs. per acre causes soil bacteria to multiply and seize 30 lbs. of nitrogen from the air, adding it to the soil fertility complex, that increment of nitrogen *may* be worth $9 (30 lbs. @ 30 cents per lb.). Also, additional phosphate, potash, and other minerals liberated from soil particles by the bacteria *may* be worth another $1, making a total of $10 for an input of only 500 lbs. of compost. For a whole ton of compost, this might amount to $40 in value $(2,000 \div 500 \text{ lbs.} = 4 \times \$10 = \$40)$.

Then, adding this $40 to an old petro-age NPK value of $15 or $20 per ton *may* equal $55 or $60—enough to create a sound market for good compost.

We are quite aware that this value concept and these particular figures may be debated. However, fellow agronomists, it is time to start debating; the gas is hissing from the gas wells and ways must be found to capture and use energy from major organic wastes.

133

Foliar Sprays

If foliar application of major plant nutrients is, in fact, a more efficient way to feed some kinds of crops, that component of value may be recognized and counted. For example, an 8-4-4 liquid fish fertilizer may be four times as efficient as a soil fertilizer in providing nitrogen, phosphate, and potash for a soybean or wheat crop. It then may have the agronomic and economic value of a 32-16-16, in dry nutrients applied to the soil.

In that case, the intrinsic and real values of good liquid fertilizers should be revised, especially when they create crop yield increases of 10 percent to 30 percent when used at only one gallon or less per acre.

Hormones Such as Cytokinins

If the English research reports are fully validated, and it is found that cytokinins in seaweed improve efficiency of photosynthesis and of carbohydrate functions in plants, then this value of marine materials may be recognized. For example, *if* a well-timed seaweed spray at one quart per acre increases the yields of sugar beets, potatoes, or grain by 15 percent or 20 percent, this creates new economic values of $25 to $100 per acre, which surely may be recognized in pricing the liquid seaweed product. *Maybe* this strategic energy-saving material is actually worth $100 per gallon. After all, aspirin sells for ten times its value based on cost. Should seaweed, too?

Seaweed, Fish and Organics for Pest Control

If preliminary research reports are validated, showing that seaweed, fish, and other full-nutrition materials may reduce pest damage to crops, this value, too, may properly be established. For example, *if* 300 lbs. of seaweed meal and two spray applications of seaweed/fish blend reduce insect populations by 50 percent in a crop, this may be worth $60 per acre, properly reflected in the prices of these products.

Even more important as time goes on, their use may reduce the need for petroleum in farming and gardening, since 90 percent of all U.S. pesticides are made from petro materials.

The above values and others like them are real. As they emerge on farms and in gardens of America, petro-paralyzed farm research leaders will need to see them, catch up, and develop new systems for measuring values of energy-saving fertilizers.

Which brings us to "organiform," an example of an organic waste product upgraded in value so that it may be shipped economically for 1,000 miles.

"ORGANIFORM"—A VERSATILE ENERGY-SAVER

One spring day in 1955 we were designing fertilizers for Seattle's Sandpoint Golf Course when a pleasant-looking man walked in and said, "I'm Jim O'Donnell. Did you want to see me?"

"Sure thing," we said, "you are just in time." A month earlier we had written to this scientist-businessman about his Nitroform, the new 38 percent nitrogen product that imitated Nature by decomposing slowly in the soil.

Son of an old New England industrial family, Dr. Jim O'Donnell was trained as an organic chemist, specializing in nitrogenous materials, and was part of the USDA research team that developed ureaform, the 38 percent nitrogen product used to fertilize lawns. But in 1953, Jim O'Donnell went into business for himself, establishing the Nitroform Company to make and sell ureaform products.

On that same spring day, our group in the Chas. H. Lilly Company agreed to serve as O'Donnell's Northwest distributor and technical-service representative. Perhaps we helped to popularize the Nitroform products and build their large markets. At any rate, Hercules Corporation of Wilmington purchased the Nitroform Company in 1959, temporarily removing the creative O'Donnell group from the fertilizer manufacturing business.

However, in 1963, Jim O'Donnell resigned from Hercules in order to reenter the field of research. It was at this time that he developed the "organiform" technology (U.S. Patent 3,942,970), and like a phoenix, soon rose again as head of Organics, Inc., based in Slatersville, Rhode Island. The firm used this ingenious process to convert leather-industry wastes into 24 percent nitrogen fertilizer products.

The "Organiform" Process

This process, in effect, *reacts* the protein of organic wastes with ureaform to make end products of substantially higher value. Amazingly versatile, it can be utilized to process and upgrade chicken litter, feathers, feedlot manure, cottonseed meal, leather scraps, sewage sludge, fish wastes, trash fish, or any other major waste containing protein. Also, it can be utilized to produce either liquid or solid fertilizers; or feeds as well as fertilizer materials.

The key is protein. Wherever there is protein in a waste, this technology can be used; and protein, being one-sixth nitrogen, is the "father" of nitrogenous fertilizers.

In its use, the "organiform" process sterilizes wastes, effectively killing and/or inhibiting pathogens and deodorizing the product. O'Donnell says, "The powerful bactericidal properties of formaldehyde are used. However, its pungent odor is not present. The ability of 'organiform' to stimulate soil microorganisms without adding harmful pathogens is considered a major improvement in the field of natural organic fertilization."

Currently, Organics, Inc. has two successful nonsubsidized plants in operation: at Slatersville, Rhode Island, making "organiform" from leather wastes, and at Winston-Salem, North Carolina, converting that city's sludge and other wastes into a 20 percent nitrogen clean pelleted fertilizer product.

Urea and formaldehyde, the key ingredients of ureaform, are utilized in making "organiform"; therefore this process utilizes *some* petrochemical nitrogen in its operations. However, the amount can be varied and regulated, depending on the availability and cost of urea.

For example, the Winston-Salem product, made from

136

sewage wastes, now containing 20 percent nitrogen can be reduced to a 12 percent level if and when urea costs and supplies get difficult.

"Organiform's" signal achievements are these:

- It effectively cycles *all* of Winston-Salem's sewage wastes back into agriculture/horticulture, capturing their energy supplies.
- It sterilizes them, achieving safety in an ecological sense.
- It upgrades the value of the waste product so it can be shipped economically into a wide area, even to Western U.S. and Canadian markets.

Since the "organiform" process is equally effective in capturing protein from wastes for feeds, we believe fishery and feedlot wastes will eventually be processed in this fashion, recovering feed protein as well as making nitrogen fertilizers. Even Carpole's, in Minnesota, may use the "organiform" process to extract the feed and fertilizer values from that region's carp and other unwanted fish.

Commenting on the waste disposal problem, Jim O'Donnell recently said, "We have millions of tons of nitrogen locked up in these wastes, but cannot use it because of old-fashioned economics and technologies. You simply cannot spend 40 dollars per ton to ship a fertilizer that is valued at only 25 dollars.

"However, if we use known skills and methods to upgrade the value of the wastes, they can be shipped economically for 500 miles, even 1,000 miles, to distant markets.

"But one thing is badly needed. The Department of Agriculture should help to educate and develop the markets for these energy-saving fertilizers, just as it does, via the Extension Service, for petrochemical fertilizers and pesticides. Then production volume can be achieved, costs can come down, and business people can make money serving the public interest.

"To date," says Dr. Jim, "we haven't received a cent of public subsidy or assistance in developing the 'organiform' process."

DR. PARR'S VENTILATED COMPOST PILES

A big problem in trying to process or recycle some kinds of wastes is their density—a lack of air in the nitrogenous material, for example in sewage sludge or fish wastes. As received from a municipal sanitation plant, sludge is wet, gooey, goopy stuff fit to make a preacher swear, to say nothing of a farmer who tries to use it for fertilizer.

Concerning this, USDA has a bright spot on its banner these days, namely Dr. Jim Parr's ventilated compost project at the Beltsville Agricultural Research Center, near Washington, D.C.

Said Dr. Parr and his associates, "Let's mix something rough and trashy with the sludge . . . something like wood chips . . . and see how that works."

In their composting field, which we visited recently, Dr. Parr mixes 3 parts wood chips with 1 part of sewage sludge, using earth-moving machinery to make and mix the compost piles. The chips aerate the mass, enabling air to enter and move. The composting bacteria may then breathe and digest the sludge, as well as some wood chips.

After three weeks in the first piles, the whole mass is turned, and given four more weeks to "cook" and cure; a total of seven weeks for the composting process.

Then comes the harvest. The piles are screened, removing the wood chips, and black earth-smelling humus emerges as the salable product. What happens to the wood chips? They go back into new compost piles, to ventilate more sludge, and slowly disintegrating eventually join the sludge as a farm and nursery fertilizer.

Asked what he thought of all this, Dr. Parr said, "It costs us from 25 dollars to 40 dollars per ton to make this compost, so we still have a cost problem. However, this is good compost, readily usable in many kinds of farming and gardening and it surely has a place in the waste recovery picture."

He then told us of his other major problem: municipal sewage systems receive all kinds of wastes, from industries as well as households, and there is a danger of contamination of the sludge by arsenic, chrome, lead, mercury, and other toxic elements not wanted in fertilizers. However, in the future,

138

this phase can be handled. Wastes can be separated and the unwanted portions kept out of the wastes that are headed for fertilizer use.

We are just beginning to mobilize skills and controls in the fields of organic waste recovery, and recycling back into the Nation's farming and gardening systems.

AND NOW . . . THE BOTTOM LINE

In chapter 1, "Pork Chops and Natural Gas," we measured America's current needs for nitrogen fertilizer at about 9 million tons per year, and found that about 3 million tons might be recovered from farm and municipal wastes.

In chapter 6, on Carpole, we discovered another nitrogen supply in carp and other unwanted fish from lakes and streams; and we estimated that the Lake States region might yield 100,000 tons of these fish-waste materials annually. However, we did not include catfish, shellfish, and ocean fish wastes, including "trash" fish of unmarketable kinds caught by fisherman but discarded.

A preliminary assessment of these additional fish resources, using data of the National Marine Fishery Service, indicates that another 100,000 tons may be available annually in accessible supplies, without encroaching on existing fish meal and pet food industries.

Added to the Lake States resource, it makes 200,000 tons of these fish materials; enough to yield about 80 million gallons per year of Carpole-type liquid fish fertilizer; enough to fertilize about 80 million acres as a nutritional supplement.

Now, using the energy-saving principles described above, we have prepared an "Energy-Saving Balance Sheet" (shown on p. 140) to illustrate the kinds of changes that may be made in providing nitrogen fertilizer for farms and gardens *as the supplies of petrochemical nitrogen are reduced by 40 percent by 1985.*

We concede that these estimates are preliminary; they may be challenged. However, it is better to be challenged in this field of public concern than to ignore the problem.

America has an amazingly high achievement in crop

ENERGY SAVING BALANCE SHEET—1985 IN TERMS OF NITROGEN FOR FARMING AND GARDENING

NITROGEN FERTILIZATION LEVEL TO BE SUSTAINED, BASED ON PURCHASES BY FARMERS, NURSERYMEN, AND GARDENERS, 1976–1977	(ACTUAL) 9.0 MILLION TONS
To be derived from petrochemicals, still, in 1985 (40 percent reduction)	5.4 million tons
To be derived directly from major farm and municipal wastes successfully recycled	1.0 million tons
To be derived from the air via increased action of soil bacteria in 60 million acres @ twenty lbs. per acre	.6 million tons
To be derived from increased efficiency in fertilizer use from foliar feeding	.5 million tons
To be derived from use of seaweed, fish, crop rotations, sap- and tissue-testing, controlled release fertilizers, hormone effects and other energy-saving materials and technologies	1.5 million tons
Total Nitrogen Uses and Alternatives, 1985	9.0 million tons

yields and food production, sustained in part by use of large quantities of nitrogen fertilizer. We do not propose to produce less, to eat less food, or to use less nitrogen.

This is the bottom line: 9 million tons of nitrogen and plenty of food in 1985. However, this achievement is possible only if we let Nature's nitrogen cycle work once more. Successful farmers will harness soil bacteria and produce their own nitrogen on their farms. Then, using foliar sprays, compost, "organiform," ventilated compost, seaweed, fish, crop rotations, and a dozen new ways to be old-fashioned, they will release about 4½ million tons of petrochemical energy for other uses.

Why?

Because it will be too expensive to use precious gas and oil crudely, grossly, and inefficiently in agriculture.

FARMING AND GARDENING IN THE TRANSITION

It has already begun. The move away from large-scale use of fossil energy in U.S. farming and gardening has begun, and it is gaining speed.

Observing this trend, *Acres, U.S.A.—a Voice for Ecoagriculture* reports that about 5 million acres of commercial farming are already under new Energy Age technologies in one form or another in the 1976 season. Charles Walters, Jr., editor-publisher, says: "You can see it happening all around, in Idaho, Texas, California, the Dakotas, Colorado, the whole Midwest. Tired of paying big fertilizer and pesticide bills and getting more problems, these farmers are rediscovering the Nitrogen Cycle. They are beginning to 'grow' their own nitrogen again.

"And they are doing it in many ways for many crops and areas: computerized farming in Idaho, Fletcher Sims compost in Texas, use of 'clod buster' humates from New Mexico, rotating the crops again, and spraying with seaweed and fish to supplement dry fertilizers put on the soil. Several million acres are being handled that way this year, increasing as farmers talk to each other and see one another's fields. Well, wouldn't you be inclined to follow the example if you saw your neighbor save $2,000 on his fertilizer and pesticide bill?

"I'll predict this," Charles Walters says. "If nitrogen fertilizers rise 100 percent again as they did for a few months in 1974–1975, you'll see 80 million acres of Midwest agriculture using leaf-spray supplements to feed the crops. And, you'll see seaweed and fish in half of the liquid fertilizers because their 'unknown factors' are giving dramatic results, increasing yields and reducing the amounts of conventional fertilizers needed."

Established in 1972 by Charles Walters, Jr., one of America's great ruralists, *Acres, U.S.A.* is rapidly emerging as the authentic voice of farmers of the future, reporting their methods and successes; telling how to do it. Its address is shown in chapter 10.

Typical energy-saving farmers in 1985 may use some or all of the following methods and materials in growing their crops:

- seed treated with liquid seaweed and fish to improve strength and viability of seedlings, giving the plants a head start, thereby increasing yields
- compost at 200 to 500 pounds per acre, or more, to encourage and feed soil bacteria
- humates at 500 to 1,000 pounds per acre, or more, for similar effects on soil bacteria
- sap- and tissue-testing of crops to adjust and improve fertilizing programs, saving on amounts of nitrogen, phosphate, and potash needed in combination with soil-supplied nutrients
- magnesium in the fertilizers to improve assimilation of phosphate and potash, thereby helping to balance the nutrients and save on purchased nitrogen
- slow-release nitrogen combined with major organic wastes, available through subsidized farm programs, in order to get the wastes back into the farming and gardening systems
- chelated and fritted trace elements in many fertilizers, since these minerals help plants to utilize nitrogen and phosphate more efficiently

142

- dry seaweed in soil fertilizers to nourish and encourage soil bacteria, and to reduce needs for toxic pesticides
- liquid seaweed and fish for use in foliar sprays and irrigation water to provide nutrients, cytokinins, trace minerals and chelating effects
- crop rotations and cover crops to "harvest" crop wastes and new fixed nitrogen provided by nodule bacteria of legumes

By using these and related methods and materials, a typical farmer may save from 50 percent to 80 percent on petrochemical nitrogen needs; and by saving both money and energy, become a patriotic businessman. A blessed event!

Home gardeners will surely resort to many of the same materials and methods. What else can they do if the oil company fertilizers get scarce and begin to cost like gold dust?

This, mind you, is only a rough description of what may occur by 1985, a year of transition, on the way to 2025, when gas and oil for agriculture will be memories, like eagles and buffaloes.

How will farmers do the job in 2025? We do not know, but can firmly predict: they will harness and use Nature's Nitrogen Cycle again, full tilt in 50 different ways. And, belatedly, a statue may be dedicated in Washington—if Washington is still the capital in 2025—to J. I. Rodale, father of America's organic movement, with the inscription: HE SAVED US FROM STARVING WHEN THE GAS WELLS WENT DRY.

HOW TO BE AS SMART AS A MOUSE WITH POTATOES

Using skills we lost ages ago, mice are fine judges of potatoes. They know which part usually tastes best, the butt end, and left to themselves in a storage cellar, they will usually nibble that end. Why?

It's this way.

As potato tubers mature, they fill the cells with starch, and the relative quantities of starch and sugar change. For example, a new potato may contain 90 percent water and only 7 percent starch, while a mature one may have 15 percent or 18 percent starch and less water; and a prime candidate for french fries or chips may have 22 percent starch. Also, as the starch rises, the weight of the potato changes, since starch is heavier than water. It weighs 1.5 grams per cubic centimeter, while water weighs only 1 gram.

This little difference gives us a tool for measuring quality in potatoes so we may be as smart as mice, who never went to school.

Going a step further, each potato has a *stem end* where it is attached to the vine, and a *butt end* where it has many eyes. As the potato matures, it tends to pack starch in the cells at the butt end first, finishing the job at the stem end. This enables us to tell whether the plants ran out of mineral nutrients to use along with starch in making the mature tubers. If so, the butt end will be heavier per cubic centimeter than the stem end.

To measure all this, unless you are as smart as a mouse, the following method may be used:

Get a snelled fishhook, a pan of water, and a sensitive spring scale graduated to 250 grams; the kind to hang on a nail with a hook at the bottom. Then, wash and dry the potato and cut it in half, separating the stem end from the butt end; and determine the specific gravity of each end in the following manner.

Put one-half on the fishhook, hang it on the scale and weigh it in air. Record this weight. Then, raise the pan of water under the potato until it is completely submerged. Record that weight. Subtract the weight in water from the weight in air, to get the weight of water displaced by this piece of potato; and, divide the weight in air by this difference to determine the specific gravity.

Repeat this weighing process with the other end of the potato, and determine the difference in specific gravity, if any. Examples:

	BUTT END	STEM END
Weight in air	321 grams	288 grams
Weight in water	7 grams	4 grams
Difference	314 grams	284 grams
Specific gravity =	$\frac{321}{314} = 1.022$	$\frac{288}{204} = 1.014$

With a difference of nearly .01, this is not an excellent potato for making chips or french fries, or for baking. It will not keep very well in storage, and the mice will surely nibble its butt end rather than the stem end.

Mice love well-fertilized potatoes.

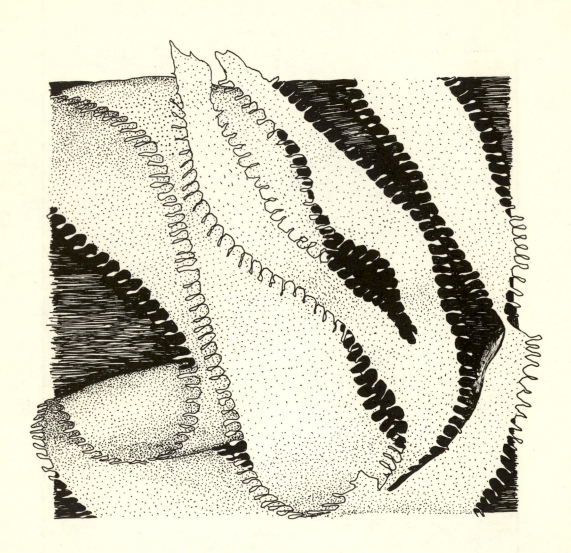

9.

RESEARCH DATA ON SEAWEED IN HORTICULTURE

Among thousands of various research projects using billions in funds, the U.S. Department of Agriculture has studied the sex life of bumblebees and flatulence in different kinds of beans, but it has not studied the values and uses of seaweed in farming and gardening. Only two Land Grant Universities have been engaged in this field of research: Clemson University and the University of Maryland; and their seaweed projects were developed with non-USDA funds. Currently, the Economic Development Administration, interested in generating marine employment for fishermen, is supporting this vital research. It is time for the landlocked and petroleum-oriented USDA to seek out and get a piece of this action.

Known worldwide for its usefulness, the seaweed research program at Clemson University was established in 1959 by Dr. T. L. Senn, using small grants from the Sea Born Corporation and assistance from the State of South Carolina, a seashore state.

Head of the Horticulture Department, Senn himself is a study in civic service not often found in research fields. To him, the world is a botanical laboratory in which he would love to work and study for five hundred years, and his research projects have a refreshing diversity. For example,

147

Senn built the first on-campus horticultural garden for blind people, a beautiful 3-acre place with rope-marked pathways, and plant markers in Braille. Run by a young blind horticultural student, this park area serves hundreds of blind people who may be seen "reading" the plant descriptions and feeling, smelling, and sensing the trees and plants. Although he is a sighted person himself, Senn produced in Braille the horticultural handbook to go along with Clemson's blind people's garden.

Also, T. L. Senn is a national leader in the use of horticulture as a resource for rehabilitation of young ex-convicts, released to his department under agreements with South Carolina's penal institutions. These young people are given an opportunity to readjust their lives and improve their personal value systems in the presence of trees, plants, and crops, while learning useful skills in horticulture.

With such wide interests, it is no accident that Senn looked on South Carolina's big seashore for seaweed and fish materials to use in growing farm crops. He started with studies of beneficial effects of seaweed on germination of seeds and the viability of new seedlings.

EARLIER WORK AT CLEMSON 1960—1973

Commenting on the seaweed research program in 1971, Dr. Senn said:

The variety of plants in the sea covers a great spectrum from microscopic species to the giant kelp. Research involving plants harvested from the sea is worldwide. However, the majority of the research is directed toward nonagricultural applications.

Liquid seaweed products were introduced commercially as early as 1950. The most commonly used is the brown seaweed, *Ascophyllum nodosum*—order Fucaceae. A British patent for a liquid seaweed extract was made as early as 1912. The nutritional value of these products does not seem to be related

148

to their nitrogen, phosphorus, and potassium content.

Their chelating activity was reported in 1959. Seaweed extracts are mostly applied as foliar sprays, and this was contrary to all concepts of plant nutrition when they were introduced in 1950. Research in the 1950–60's proved that foliar feeding was effective. Foliar feeding became orthodox practice in the 1960's. This notable change helped the sale of liquid seaweed products throughout the world.

Research with various extracts from *Ascophyllum nodosum* was initiated at Clemson in 1959. Greenhouse and field studies with horticultural plants sprayed with seaweed extracts exhibited growth patterns that seemed to be induced by the presence of growth regulators in the seaweed extracts used.

At present, growth promoting substances in seaweeds are well known. It has been shown that rice seeds germinated more readily when soaked in extracts of a blue-green alga. Aqueous extracts have proven more effective than alcohol extracts. Two papers in 1965 and 1967 by Schiewer extend the range of seaweeds which contain indole compounds. He stated that alkaline hydrolysis released large amounts of auxins. Various reports have been made on the presence of gibberellins.

Much of the research in the United States related to the agricultural uses of seaweed has been conducted at Clemson University. It has shown that seed germination of some plant species is increased by treatment with seaweed extracts. Treatment of seeds of several species with seaweed extracts resulted in greatly accelerated respiratory activity. As the concentrations of seaweed extract were increased, the rate of respiratory activities also increased. At the highest concentrations studied, where the respiratory activity was highest, seed

149

germination was low. Changes in respiratory activities of plants have also been induced through applications of seaweed meal and extracts. These results suggest the presence of a respiratory stimulant in the seaweed meal and extract. This unknown material may be a hormone commonly found in plants, or only in marine vegetation.

Fruit harvested from peach trees which had been sprayed with seaweed extracts had increased shelf-life when compared to similar fruit from untreated trees. The growth of decay organisms, which may be present on the surface of the fruit during the growing season and following harvest, appeared to be inhibited by treatment with seaweed extracts. Earlier sprays, beginning during bloom, resulted in the best fruit shelf-life.

From previous field studies, the changes in plant growth patterns seem to be induced by the presence of growth regulator(s) in the seaweed extracts employed.

Several pieces of evidence suggest the presence of a growth hormone of some type in the seaweed extracts. More evidence will be required to definitely say which one(s) is (are) involved. Explanation of the observed responses will depend on the type or class of hormone determined.

The occurrence of auxins and gibberellins in fresh algae has been studied. However, no detailed quantitative measurements were made and no work has been reported on seaweed extracts. Whether these regulators survive the processing method is of interest because the use of fresh seaweed is quite limited, but the use and potential uses of seaweed extracts, be it in the meal form or liquid form, are almost limitless.

The existence of unknown growth regulators in seaweed extracts is another possibility. These unknown regulators can be an entirely new class of regulators, such as those recently discovered from grape pollen, or be modified compounds of known regulators. The isolation and identification of either would be of particular interest to plant science and industry.

Blunden and Woods studied the effects of various carbohydrates in seaweed fertilizers on plant growth. Various effects were demonstrated which suggest that certain sugars present in the extract may serve as additional energy sources for plant growth.

A complete analysis of the seaweed extract should be conducted. In addition to the analyses of all organic matters, carbohydrates, proteins, lipids, and pigments, the chemical identification and quantitation of plant growth regulators shall be the main part of the composition analyses.[1]

Illustrating the significant effects of seaweed on respiration (biological activity) in seeds and plants, Senn provided, among other units, the following research data: [2]

EFFECTS OF SEAWEED EXTRACT ON SUBSEQUENT RESPIRATORY ACTIVITY OF LOBLOLLY PINE SEED, 1960

TREATMENT	CUBIC MM. CO_2 EVALUED PER GRAMS PER HOUR
Check	103.5
1 part to 100 H_2O	193.1
1 part to 25 H_2O	258.5
1 part to 5 H_2O	369.9

[1] Senn, Skelton, Martin, and Jen, South Carolina Agricultural Experiment Station, *Seaweed Research at Clemson University 1961–1971.* Research Series No. 141 (Clemson, South Carolina: May 1972).
[2] See previous citation.

EFFECT OF SOIL APPLICATIONS OF SEAWEED MEAL
ON RESPIRATORY ACTIVITY OF GERANIUM LEAVES

TREATMENT	CUBIC MM. CO_2 EVALUED PER GRAMS PER HOUR
Check	287
5 percent seaweed in soil	374
20 percent seaweed in soil	409
10 percent seaweed in soil	440
40 percent seaweed in soil	482

Moving from the pioneer work on respiration of seeds and plants as affected by seaweed, the Clemson group then studied specific benefits from seed treatments, such as soaking beet seed in dilute seaweed extract before planting. In this case, germination of the treated seed was 84 percent after one week, while untreated beet seed was zero, indicating that seaweed speeded up the germination process. Further studies were made of many kinds of seeds, including zinnia, tobacco, radish, peas, turnips, tomato, cotton, pine, and holly.

Summarizing this phase, Senn said, "In all cases tested, application of the extract, even in low concentrations, increased the respiratory activity of the seed. . . . The optimum concentration varied considerably with different species of seed. Seaweed extract appears to contain the potential to increase germination of certain seed." [3]

This early work by Senn and his group on seed germination was undervalued by many observers. They said, "Why try to increase germination of seeds? U.S. seeds have high quality and a high rate of germination anyway. Why try to improve it?"

However, the Clemson research was showing: (a) seaweed significantly stimulates plant life, and (b) treatments with seaweed can help growers to cope with germination problems in cold soils and wherever a stimulus is needed.

[3] T. L. Senn and R. J. Skelton, *Review of Seaweed Research—1958–1965*. Research Series No. 76. (Clemson, S. C.: South Carolina Agricultural Experiment Stations, Feb. 1966).

EFFECTS ON QUALITY AND
SHELF LIFE OF PEACHES

During earlier work at Clemson, the research workers observed that molds were inhibited in soils mixed with seaweed when stored in barrels for the respiration and other studies. Also, European researchers had reported an effect of seaweed in controlling molds and fungi. Senn therefore postulated, "Perhaps seaweed will help our peach-growers to control mold and rotting of fruit"; and he launched the peach studies in 1965. The design of this experiment follows:

An orchard of Harvest Gold peaches located in Cherokee County, South Carolina, was selected for the experiment. Treatments used were: check, seaweed meal, two seaweed sprays, three seaweed sprays, four seaweed sprays, and five seaweed sprays, with and without seaweed meal. Each treatment was replicated six times. Where meal was used, it was applied broadcast in the fall at a rate of two pounds per tree. Seaweed sprays were prepared by mixing one gallon of seaweed extract per 100 gallons of water and applied to the tree until drip-off. Where more than one spray was applied, the applications were every two weeks, starting June 8, 1965. All treatments received the standard peach spray program.

Fruit samples were harvested August 29 and returned to the Horticulture Laboratories of Clemson University for analyses. Determinations were made for firmness, soluble solids, total titratable acidity and pH on the fruit samples. Standard analytical procedures were used. Additional fruit were used for shelf-life studies.[4]

The results of these various seaweed applications on shelf life of peaches were summarized in this report:

[4] R. J. Skelton and T. L. Senn, *Effect of Seaweed on Quality and Shelf-Life of Harvest Gold and Jerseyland Peaches*. Research Series No. 86. (Clemson, S.C.: South Carolina Agricultural Experiment Station, Oct. 1966).

153

EFFECT OF VARIOUS COMBINATIONS OF SEAWEED MEAL AND EXTRACT ON SHELF-LIFE OF HARVEST GOLD PEACHES, 1965.[5]

TREATMENT *	PERCENTAGE OF PEACHES ROTTED (DAYS AFTER HARVEST)			
	7	14	21	28
Check	5.6	13.3	20.0	43.3
Check and meal	1.7	7.2	12.8	23.9
Two sprays	4.4	8.9	17.8	28.9
Two sprays and meal	2.8	9.4	14.4	34.4
Three sprays	1.7	5.6	10.6	25.6
Three sprays and meal	5.0	14.4	27.2	43.9
Four sprays	4.4	7.2	11.1	31.7
Four sprays and meal	4.4	13.9	21.1	31.7
Five sprays	3.9	7.2	13.3	29.4
Five sprays and meal	7.2	17.8	28.3	46.1

[5] See previous citation
* All trees had a regular peach spray schedule.

It was noted that moderate seaweed spray applications seemed to give optimum effects. Summarizing this work, Drs. Senn and Skelton said, "Application of seaweed sprays resulted in increased shelf-life of peach fruit." They attributed this effect to seaweed's ability to inhibit growth of decay organisms.

FURTHER SEAWEED STUDIES—1965–1973

With support from the National Science Foundation, the Sea Born Corporation, and the State of South Carolina, the Clemson University seaweed research program was sustained during 1965–1973 despite the absence of Agriculture Department funds for this purpose. The work on shelf life and quality of fruit crops was extended to include tomatoes. Also, work was done in 1969–1970 on seaweed fertilization of camellias, coleus, orchids, tobacco, and vegetable crops. The effects were generally significant and beneficial, as indicated by Senn's summary at the close of this section.

Meanwhile, Senn and Skelton and others of the staff turned their attention to the chemical and hormonal composition of seaweed, endeavoring to isolate the "unknown growth factors" that caused plant growth which could not be accounted for by accepted principles of plant nutrition.

When we consulted with Senn in 1972, he said, "I am strongly in favor of studying pest control and many other practical uses of seaweed in horticulture. However, we still need more basic research to clear up mysteries as to how and why this marine material can do what it clearly is able to do, stimulating growth and regulating biological processes in plants. Especially, we need to study hormone factors and how they work."

CLEMSON SEAWEED RESEARCH—1973–1976

Sustained by private grants * and the State of South Carolina in 1973–1974, Clemson's seaweed research program

* These included Maxicrop, U.S.A., headed by Per Bye Ohrstrum, a well-known figure in U.S. seaweed affairs.

won Economics Development Administration financing for 1975–1976, due to its evident ability to stimulate new employment in coastal districts where fishermen and fishery workers might use their skills and equipment harvesting and processing seaweed and fish for agricultural uses. Therefore, this valuable research program has a new life and outlook.

In a basic experiment launched in 1974, liquefied seaweed was applied to side-by-side plots planted as follows:

CONTROL	SEAWEED SPRAY *
Tomato	Tomato
Okra	Okra
Pea	Pea
Soybean	Soybean
Corn	Corn

* Seaweed spray applied at 100 parts per million weekly.

Among reports of results, the following are noteworthy:

1. INFLUENCE OF *ASCOPHYLLUM NODOSUM* SPRAYS ON YIELD OF SOYBEANS *

TREATMENT	ESTIMATED YIELD	
	NH$_3$	PROTEIN
Control	9.28	47.70
Seaweed spray	9.90	50.89

* The gross yield of crop would be much larger.

2. INFLUENCE OF SEAWEED SPRAYS ON NODULATION OF SOYBEANS

TREATMENT	MEAN COUNT
Control	96.0
Seaweed spray	102.5

156

3. INFLUENCE OF SEAWEED SPRAYS ON YIELD OF OKRA

TREATMENT	WEIGHT	NUMBER OF PODS	SIZE
Control	7483	412	18.16
Seaweed spray	9469	487	19.61

Note: Some of the above results are in terms of mathematical "means," using Duncan's Multiple Range Test.

These studies in 1974–1975 include many other crops, such as sweet corn, tomatoes, turnips and broccoli. Also, a major research unit is being conducted on pecan trees to determine the capability of seaweed in preventing tree decline, and in rehabilitating run-down pecan groves.

Summarizing seaweed research at Clemson University during 1959–1975, T. L. Senn says:

The chelating activity of *Ascophyllum nodosum* order *Fucaceae* was reported in 1959.

Ascophyllum nodosum extracts contain *phenols* which aid in root development and exhibit growth-promoting properties.

The magnesium content of peach leaves treated with seaweed extracts was higher than nontreated plants.

Apparently seaweed extracts can supply enough magnesium, manganese, zinc, and boron for pine-apple seedlings (*Citrus sinensis var.*) to carry on near-normal respiration.

Seaweed extracts improve plant utilization of boron, copper, iron, manganese, and zinc. The improved utilization is due either to chelation or improved metabolic activity.

Treatment of seeds of some species of plants with seaweed increases germination and results in

157

greatly accelerated respiratory activity. As the concentrations of seaweed extract increase, the rate of respiratory activities also increases.

These results suggest the presence of a respiratory stimulant in seaweed meal and extract.

Fruit harvested from peach trees which had been sprayed with seaweed extracts had increased shelf-life when compared to similar fruit from untreated trees.

The growth of decay organisms which may be present on the surface of fruit during the growing season and following harvest appear to be inhibited by treatment with seaweed extracts.

Solutions of Norwegian dehydrated seaweed extract from *Ascophyllum nodosum* sprayed at weekly intervals onto cucumbers during fruiting increased yield.

The growth promoting substances in fresh algae have been studied and well-documented. Recent research at Clemson University suggests the presence of an auxinlike, gibberellinlike, and cytokininlike hormone in the seaweed extracts.

The benefit of seaweed extracts could arise from physiological changes in the internal structure of plant cells which could be induced by the high contents of the micro-nutrients in the seaweed extracts.

One year's research results at Clemson University showed a significant increase over the controls in soluble solids of field grown tomato, in protein content of soybean, and in insoluble solids of Concord grape when plants were sprayed with extracts of *Ascophyllum nodosum* (weekly sprays of 100 ppm, 100ppm, and 20,000 ppm respectively).

It would be well worth the effort and financial investment to further explore the composition, bio-

158

logical activity, and nutrient components of seaweed extracts.

Based on modern scientific technology and public concern for ecology, the marketing trend is definitely in the direction of nontoxic, nonpolluting organic products. Seaweed derivatives definitely offer ingredients that could feasibly be incorporated into such products.[6]

Among the many achievements in Senn's career, his successful leadership of America's seaweed research, virtually unaided, will stand as one of the finest.

SEAWEED RESEARCH AT THE UNIVERSITY OF MARYLAND

It is the view of the Environmental Protection Agency that research evidence of ability of any material to control insects or other pests should be developed, if possible, at more than one Land Grant University. Therefore, the seaweed research program was expanded in 1973 to include the University of Maryland. Under inter-university agreements, funds provided to Clemson have been shared by the Agronomy Department at Maryland and have been utilized for specific purposes in 1973–1976. This phase has been handled by Dr. James Miller, Chairman of Agronomy, and Dr. John Hall, agronomist.

Commencing with Clemson's informal observations that seaweed may assist plants in resisting pest damage, it was decided that the Maryland work would be addressed to two sectors: (a) possible control of Mexican spotted beetles in soybeans, and (b) possible control of nematodes and fungus diseases in turf. However, various factors have prevented development of the soybean unit, and turf pest control with seaweed has been the principal research interest at Maryland.

Putting this project in perspective, nematodes, as many

---

[6] T. L. Senn, unpublished statement January, 1976.

FINDING THE FAIRBANKS FACTOR IN POTATOES

In 1956, Alaskan farmers grew potatoes, grain, silage, and vegetable crops successfully but were handicapped because they used temperature-zone fertilizers in a sub-Arctic climate. And the worst problem was a catastrophic collapse of many potato fields in late July. For unknown reasons, the vines just collapsed and died.

As their fertilizer suppliers, we felt a responsibility and employed Dr. George Scarseth, then head of American Farm Research Association, to help in overhauling Alaskan fertilizers and solving this costly problem. It was in the era of Drs. Allen Mick, Arthur Buswell, and Curt Dearborn, superb Agricultural Experiment Station leaders.

Scarseth called it the "Fairbanks factor," because the "disease" was so unique in the Fairbanks vicinity, showing as a funny chlorosis of the leaves in late July. Scientists came from thousands of miles away to study this "disease." Even Dr. Dixon, renowned plant pathologist, came to look at it. But no one could answer the question: "Where does 'hunger' end and 'disease' begin? We saw the Fairbanks factor as a special urgent hunger, like that which diabetics have for sugar when their blood supply runs low.

Our secret weapon, in this case, was sap testing. If the mineral isn't in the sap at the critical time when it is needed, the plant is starved, no matter how much is in the soil or fertilizer. Analysis of the sap may give the accurate clue.

In late July 1956, we ran sap tests with Dr. George Scarseth on potato fields in the Tanana Valley near Fairbanks. And lo! there was near-zero potassium. How could this be, in soils with 400 lbs. per acre of available K_2O, and in fields fertilized with 1,000 lbs. per acre of a 10-20-10 potato fertilizer? But it was true. Sap tests don't lie.

Here was the problem, defying conventional research: these potato farmers banded their fertilizer at planting time, alongside the rows, and the potatoes grew gloriously. Then, in early July, they hilled their potatoes to cultivate and control weeds, then left them alone. But with 22 hours per day of Arctic sunshine, the potatoes tuberized like mad day and night, demanding potash. They just plain ran out of potash, collapsed, and died.

On digging up hills of potatoes we found the plants had filled the unfertilized hilled soils with roots, leaving the banded 10-20-10 unused below; and the new roots were unable to gather potash from the unfertilized soil fast enough to meet the demands of the plants.

That's often the story of farm research. If you use old guides, you sometimes fail. Use new ones and you often fail. Use combinations of the old and new, and sometimes you win. But it is as hard to be a hero in farm research as anywhere else today.

readers know, are tiny wormlike organisms that feed on plant roots, causing lesions and mild to severe damage to plants and crops. Overall in the U.S., nematodes curtail yields by approximately 10 percent, causing massive damage dollarwise in vegetable, fruit, and field crops. The nematode damage to soybeans, alone, is estimated to be $800 million per year.[7]

In the case of turf grass, nematode infestation is linked to occurrence and control of *Fusarium roseum*, a highly pathogenic disease of Kentucky bluegrass. Where nematodes prosper, this fungus disease often thrives, and controlling certain turf nematodes appears to have the potential of reducing injury from *Fusarium roseum*.

THE FIELD TRIALS WITH SEAWEED

Utilizing field plots 20×20 feet in size, Dr. Hall established in 1973 replicated test applications comparing seaweed with commercially available fungicides and nematicides. The seaweed applications included seaweed meal at the rate of 250 lbs. per acre and liquefied seaweed at 1 gallon per acre. Then, at predetermined intervals, the numbers of nematodes per 200 grams of soil were counted.

As expected, not much effect was attained in the seaweed plots the first year, since the soil environment and food chains had to be modified with seaweed materials, a slow but relatively permanent process. However, a modest reduction in certain nematode populations was observed in 1974, and a statistically significant effect was attained in 1975.

Informally reported, the nematode counts show the following results in one series of the test plots:

TREATMENT	PARATYLENCHUS PIN NEMATODES PER 200 GRAMS OF SOIL
Control plot	344
Nematicide	104
Seaweed meal	7

[7] These estimates are from the Plant Pathology and Nematology Section, Agricultural Research Administration, U.S. Dept. of Agriculture, Washington, D.C.

Tentatively, it is believed that the control is being generated from the seaweed meal material, rather than liquefied seaweed or a combination of the two forms of seaweed; therefore, in future work the straight meal may be suggested.

In an overall comparison of the seaweed with the commercial nematicide, these reconnaissance tests suggest that seaweed meal is about 50 percent more effective than toxic nematicides in reducing certain nematode populations in turf and can be applied at substantially less cost.

The nematode species being reduced by seaweed applications in the Maryland tests are: *Paratylenchus*—the pin nematode, and *Pratylenchus penetrons,* the lesion nematode.

Commenting on these seaweed investigations, Dr. John Hall says:

> It is much too early to assess the importance of this research on the value of seaweed to control nematodes and *Fusarium roseum.* We are taking the ecological approach involving many food chain and environmental factors, which is inherently a slow process. However, the results in this third year of working with seaweed are very promising and surely justify further investments of time and money. At present it appears that seaweed meal applied at about 250 pounds per acre may in three years' time substantially reduce certain nematode populations in Kentucky bluegrass turf.

Where else in farm and garden research is there a brighter, more interesting spot than this one at the University of Maryland, where less than $25,000 of research funds have made a beachhead for fighting America's 4 billion dollar per year nematode problem?

THE CYTOKININ FACTOR

Young, brilliant and capable, Gerry Blunden of Portsmouth Polytechnic, England, has emerged since 1970 as the world's leading investigator of the role of cytokinins in seaweed's ability to stimulate plant growth. And, as readers will remember from chapter 4, cytokinins are a hormone group that may affect photosynthesis and carbohydrate functions in plants. Relentlessly, Blunden is pursuing the relationships between seaweed, *cytokinins,* photosynthesis, carbohydrates, and crop yields.

Two units of the English research in this field are especially noteworthy, and may be useful to stimulate parallel investigations in the U.S. The first of these is a study of cytokinins, per se, reported by Brain, Chalopin, Turner, Blunden and Wildgoose.

Introducing this report, Blunden's group said:

> Beneficial results from the use of seaweed extracts as fertilizer additives have included increased crop yields, improved seed germination, increased resistance of plants to frost and to fungal and insect attack, increased uptake of inorganic constituents from the soil, and reduction in storage losses of fruit. The rate of application of seaweed extract is low in terms of solid content and hence the active compound, or compounds, must be effective in very low concentration. Cytokinins can produce some of the results claimed for seaweed extracts and Booth suggested that these compounds may be responsible for much of the activity of seaweed extracts. . . . The commercial seaweed extracts themselves do not appear to have been tested for cytokinins and, in view of the possible significance of these compounds in the extracts, a commercial aqueous seaweed extract, S.M.3 (Chase Organics), prepared from species of *Laminariaceae* and *Fucaceae,* and containing 22 percent solids, was tested.[8]

In this test, carrot tissue, belladonna, and radish leaf were grown, separately, in solutions of (a) kinetin and (b)

[8] Brain, Chalopin, Turner, Blunden, Wildgoose, "Cytokinin Activity of Commercial Aqueous Seaweed Extract," *Plant Science Letters* (Amsterdam: Elsevier Scientific Publishing Company, 1973).

164

seaweed extract, along with uniform nutrient supplementations. Growth responses of these plant tissues were then measured, and it was found that seaweed extract provided cytokinins as well as, or better than, the straight kinetin solution. The report says:

> The results of the three independent methods used for the demonstration of cytokinin activity indicate that the commercial aqueous seaweed extract contains cytokinetic material. Concentrations of 10–100 mg./1 of *cytokinin* have been used to produce beneficial changes in plants and the results from the tissue culture of *A. belladonna* and the radish leaf bioassay show that the commercial seaweed extract has a cytokinin activity which would produce such physiological changes, even when applied at the low concentrations used in practice. The cytokinin content of the extract is probably lost when applied to the soil, but seaweed extracts are generally recommended for foliar application and it is well known that *cytokinins* are readily absorbed through leaf surfaces.[9]

THE POTATO TRIALS

Dr. Gerry Blunden's potato trials were conducted in a two-year series, in 1974 and 1975, on farms near Chelmsford, Essex. Concerning these units he said:

> Many different beneficial effects have been recorded for crops treated with seaweed extracts. As well as increased crop yields, other benefits have included improved seed germination, increased uptake of inorganic constituents from soil, increased resistance of plants to frost and to fungal and insect attack and improved storage quality of fruit. Most of these effects claimed for seaweed extracts can be

[9] See previous citation

165

explained by the action of cytokinins and, in 1973, Brain and coworkers demonstrated a high cytokinin activity of a commercial seaweed extract. It was later shown that this activity was common to the most frequently used seaweed extracts on the U.K. market.

The type of response expected from a plant treated with a cytokinin suggested that carbohydrate-storage crops would be ideal for initial trial work. Trials on sugar beet showed that plants treated with seaweed extract had a higher root-sugar content than control plants and that the juice purity was higher from the treated plants because of a reduction in the content of amino nitrogen and potassium. These results showed a close correlation with those to be expected from treatment with a cytokinin. The trials did not, however, incorporate the use of a reference cytokinin. As a continuation of this work, trials on another carbohydrate-storage crop, namely potatoes, have been conducted. Blunden has recorded that potato plants treated with seaweed extract gave increased yields of tubers and that the potatoes from the treated plants were more even in size. This report, however, was based on commercial observations and not on properly conducted trials. In this present report, results obtained from seaweed-extract-treated plants are given and compared with those obtained from plants treated with the synthetic cytokinin, kinetin.[10]

In the 1974 plots, King Edward and Pentland Dell varieties were used in four replications with these basic treatments:

Kinetin * at 250 and 125 parts per million applied at rates equivalent to one gallon per acre (11.22 litres per hectare).

[10] Gerald Blunden and Paul Wildgoose, *The Cytokinin Effects of Aqueous Seaweed Extract on Potato Yields* (forthcoming).
coming).
* Kinetin is a synthetic cytokinin substance.

Seaweed extract containing 125 parts per million of cytokinin, at rates of 1 gallon and ½ gallon per acre, diluted with 100 gallons of water.

Couch grass invaded the Pentland Dell plots, invalidating that portion. The results for King Edward were as follows:

In the 1975 units, King Edward variety was used in a study of growth and yield responses to foliar applications of seaweed extract at 1 gallon per acre, used in conjunction with soil applications of a 15-15-19 potato fertilizer.

In these plots the yields from seaweed-treated plants were from 20 percent to 27 percent higher than from non-sprayed plants, with greater response when less of the 15-15-19 soil-applied fertilizer was used, indicating that a liquid seaweed spray might effectively supplement the conventional fertilizer.

Summarizing the potato investigations, Dr. Jerry Blunden said:

> The results from the 1974 potato trial show that the use, as a foliar spray, of aqueous seaweed extract with a cytokinin concentration equivalent to 125 ppm of kinetin, at a rate of application equivalent to 1 gallon per acre (11.22 litres per hectare), produces a significant increase in the yield of potatoes of the variety King Edward. A significant increase in yield was not achieved with the variety Pentland Dell. When the seaweed extract was applied at a rate equivalent to ½ gallon per acre, no significant increase in the crop of either King Edward or Pentland Dell potatoes was achieved. Application of kinetin at a rate equivalent to 1 gallon per acre of a 125 ppm solution resulted in a signific the yield of potatoes of the variety King Edward, but not of the variety Pentland Dell. There is no significant difference in the results from this kinetin application and the results from the application of the seaweed extract of equivalent cytokinin activity. This close correlation strongly indicates that the beneficial result from the use of the seaweed extract is due to its cytokinin content. Application of kinetin at a rate

167

POTATO TRIALS—1974

TREATMENT	RATE	TUBER YIELD: PERCENT OF CONTROL
Seaweed	½ gal./acre	106.1
Seaweed	1 gal./acre	111.6
Kinetin (125)	1 gal./acre	112.1
Kinetin (250)	1 gal./acre	128.9

equivalent to 1 gallon per acre of a 250 ppm solution produced highly significant increases in the yield of potatoes of both King Edward and Pentland Dell varieties.

In the trials conducted, the use of seaweed extract at the rate of either 1 or ½ gallon per acre was chosen because these are the rates of application generally used in practice. Good results were obtained by using the seaweed extract at the rate of 1 gallon per acre, but as the best result was achieved using kinetin at a rate equivalent to 1 gallon of a 250 ppm solution per acre, and as a close correlation was found between the results from plots treated with seaweed extract and kinetin of equal cytokinin activity, there is a strong indication that an application rate greater than the 1 gallon per acre of the seaweed extract used in this trial would have produced a better increase in the crop yield. Further trials are necessary to determine both the optimum rate and time of application, but the use of seaweed extracts or other cytokinin materials for increasing potato yields looks promising.

In 1975, Blunden served as a guest lecturer at the ACRES, U.S.A. National Conference in Kansas City, answering hundreds of questions and inspiring young people to remain in agriculture after all. Superseding the lesser matter of whether Jerry Blunden is right or wrong in his seaweed research, he is already serving as a beneficial force. Like a fresh ocean breeze, he is helping to clear out the musty corridors of farm research in America.

SEAWEED AND FISH RESEARCH— AN OVERALL VIEW

It would be easy at this point to castigate research leaders of the Department of Agriculture, and to aim bazooka shots at a number of leadership personnel in the Environ-

169

mental Protection Agency . . . all of whom had deaf ears when seaweed research desperately needed support in recent years. The soft clink of petrodollars for fund-matching support of farm research made better music to those people at that time.

But even more compelling, science has fads and trends like clothing, and wearing the petroleum hat was a good personal policy up to a couple of years ago if you were employed in farm research fields. Now times and hats are changing, and the energy-saving styles are better.

Soon, research in energy-saving potentials of seaweed, fish, manures, and sludge will be a roaring custom in universities and federal agencies and you will be out in the cold if you are not already in it. America smothers and kills most things it loves, often choking them to death with money. It is a poor way to die.

So, where are we now in seaweed research, and what about fish and the organic things? Are these part of the same essential technologies?

As we see it, Clemson, Maryland, Portsmouth Polytechnic, and hundreds of farmers have already answered the basic questions: "Is seaweed a strategic energy-saving fertilizer?" "Does seaweed have an important future role in protecting plants and crops against insects and plant diseases?"

Unequivocally, the answer is "yes" to both questions. Sound next steps in these research fields are to design ample basic research, and disseminate the guides and values from Clemson and Maryland into dozens of other research institutions and Land Grant Universities, so they can save time and money establishing studies for their own kinds of crops and gardens.

Collaterally, more research is needed concerning foliar feeding of plants, since many of the energy-saving materials will be applied in liquid forms. Also, leaf feeding of major NPK nutrients will conserve on supplies as these fertilizers become more expensive and scarce.

FISH AND THE CYTOKININ FACTOR

In due time, the cytokinin factor will come into focus, logically related to other nutrients for plants; however, at present, it seems a bit confusing. For example, Blunden and his group seem to be finding that cytokinins are the stellar performers, with better results at higher than gallon-per-acre applications, while Sea Born and other Midwest seaweed people are getting substantial yield increases from a quart per acre or less of non-cytokinin-rated seaweed products.

How can these differences be explained?

As we see it, there is no basic dilemma. If plants require and/or respond to cytokinin hormones, it is quite certain that they learned to synthesize these substances ages ago, in the misty origins of their evolutionary careers.

Many organic materials contain keto acids, which can serve as precursors for cytokinins. For example, fish protein has amino acids which may be reduced to keto acids via the causative processes of oxidation. Therefore, fish can provide essential building blocks for making cytokinins.

Fish hydrolysate contains no cytokinin, and liquid seaweed contains no keto acids. However, both are of marine origin and have concentrated the full range of trace minerals of seawater. Theoretically, both are able to satisfy needs of plants for cytokinins, *or* for the essential units for making these within their own life systems.

Minerals, enzymes, hormones, foods, functions: all are logically correlated as they form and serve each other.

The various organic wastes all contain amino acids (proteins) and such assortments of minerals as they were able to obtain when growing as living tissues. In turn, this depended upon their climate zones and specific environments. Fish and seaweed, growing in a mineral soup, came out best of all.

Seen in this fashion, peace reigns in the research camps . . . for a few minutes, at least. Life on earth makes and uses *everything*, or it smothers, starves, or dies. Look on the other side of a cytokinin and you will find an amino acid; on the other side of an enzyme and you will find a mineral; on the other side of a disease and you will find a hunger; on the

171

other side of a food and you will find a poison; and on the other side of death and you will find resurgent life.

Within these dichotomies the truth will be found about seaweed, fish, organic wastes, and how to fertilize farms and gardens as the gas wells die.

Seaweed is an amazingly complex and complete substance. Reports of its composition are given in the Appendix.

Sea Born

172

PRACTICAL FARM RESEARCH

The picture on page 172 shows six large heads of wheat and six smaller ones, over the words "sprayed" and "unsprayed." It is an example of practical farm research, unofficial and scarcely scientific, but honest and useful to hundreds of farmers in Arkansas.

In this case, Gerald Costner, owner-manager of Costner Farms near Manila, Arkansas, sprayed part of his 1,000 acre wheat field with Sea Born Plus F, the new blend of liquefied seaweed and fish, using only one quart per acre diluted in water. It was supplied by Harry Wright, Costner's friendly fertilizer dealer.

Three weeks later, when the pictures were taken, the wheat heads in the sprayed portion were about 25 percent larger and heavier than heads in the rest of the field, and the harvest will probably show a 20 percent to 30 percent increase in yield . . . a whopping big net profit from a $6 or $8 investment.

Already, many neighbor farmers have visited Costner's wheat field and he is, in effect, serving as their research facility and adviser. Well, suppose his tests are invalid and the guidance is erroneous, causing the neighbor farmers to lose money buying poor products? That could happen, but usually it doesn't. Leading farmers like Gerald Costner are skilled people and they seldom mislead their neighbors.

This is practical farm research, unscientific but tremendously valuable in American agriculture, coping with energy and pollution problems in the face of resistance to change found in some of the USDA research centers.

If the new seaweed-fish fertilizers give excellent results in places like Manila, Arkansas, these farmers will go to the Department of Agriculture and say: "Please test this stuff and give us research guidance for using it." The first year, the answer may be, "Sorry, we have no money or room for such a project." However, when thirty or forty farmers of the area successfully use the new product, the USDA Experiment Stations may pick it up and act as if they invented it in the first place.

Too often, that is how farm research is improved and redirected to new horizons and products. It is based initially on a local demonstration, such as Costner's wheat field sprayed with seaweed and fish.

Was this test replicated?

Come on, now, these people hardly know what that word means. Some of them may think it is a kind of a lizard, but they are good at growing wheat and they know a fine crop when they see one.

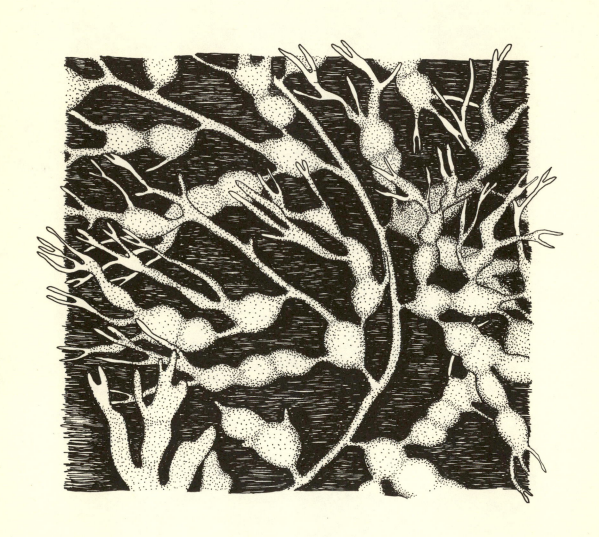

10.

THE COMPLETE MINERAL FERTILIZATION OF PLANTS

While living in the Northwestern United States and fertilizing hundreds of farms and gardens in that region—and in Alaska—we were always face-to-face with certain questions: "Are these soils fertile?" "Do they need extra minerals?" Isn't it enough to fertilize with manure, rock phosphate, compost, and lime, as we do?" These are typical questions asked by "organic" farmers.

With the tall Cascade Mountains cutting that region into two drastically different climate zones, it was a classic place for studying those questions and giving useful answers. For example, here is a comparison of the two halves of Oregon and Washington:

1. The Humid Western Part

Rainfall in this portion ranges from 30 to 180 inches (15 feet) per year and percolation of water is downward through the soil particles, carrying minerals out and away from the topsoil, moving forever to the sea. Moreover, this has been going on for centuries and geological ages.

As fast as minerals erode from country rock in Western Oregon and Washington due to freezing, thawing, bacterial effects, and other weathering actions, the seasonal rains wash

177

the unused supplies away to river deltas and the Pacific Ocean. Thus, they are lost to farming.

A typical soil analysis from this region, for example in the Chehalis or Albany area, will show only 800 or 1,000 lbs. of available calcium per acre; a lime requirement of 3 or 4 tons per acre; only 20 or 30 lbs. of phosphate—usually less; 120 lbs. of potash; very little magnesium; and hardly enough of the trace minerals to "put in your eye," as the old-timers used to say. These are *acid* soils with typical pH levels of 4.8 to 6.2.

Now, are you an "organic" farmer or gardener in this area, or in one like it in Maine, Connecticut, Wisconsin, Georgia, or Mississippi? If so, where will your adequate mineral supplies come from for mineral-rich, nutritious, pest-resistant crops? Since the shortages of minerals in the leached-out soils are also carried into animal feeds, manures, composts, crop wastes, cover crops, and worm castings, supplementary supplies must be imported or people and animals will be malnourished.

That is the basic agronomic problem of human cultures in high rainfall areas. They must import their minerals from dry country in foods and feeds, retrieve them from the sea, or buy special fertilizers.

2. The Dry Eastern Part

Rainfall in this beautiful portion of the Northwest ranges from less than 10 inches per year in the Moses Lake or Burns districts to 16 inches at Spokane. Yakima, Washington, gets along on only 8 inches of annual rainfall.

In such climate zones, soil minerals, if undisturbed by man, remain in the topsoil, or leach *upwards*, accumulating in "alkali spots" at the surface. Or, due to seasonal movements up and down with soil water, the minerals may accumulate in caliche layers within the soil; whitish hard deposits of calcium, sodium, magnesium and other earth minerals.

Typical soil analyses of such dry area soils may show 10,000 to 20,000 pounds per acre of available calcium, 200 lbs. of phosphate, 400 of potash, 300 of magnesium, and substantial latent supplies of iron, manganese, boron, zinc, copper, and other minerals, as provided in weathering of local rocks—but like pots of gold at the end of rainbows, such elements are often unavailable for use.

These are alkaline soils, of pH levels 7.0 to 9.0, often with scant organic matter, because hot summer weather breaks it down as fast as it accumulates.

Are you an "organic" farmer or gardener in this climate area, or one like it in Arizona, Texas, California, or Colorado? If so, you have other problems. The minerals tend to "tie each other up" so your plants get too much of some and too little of others; the high pH immobilizes iron, zinc, manganese, and other essential nutrients; and no matter how hard you try, it may be impossible to raise the organic level in your soil to 5 percent, because evaporation works as fast as you add manures and other farm wastes.

These are problems of human settlements in low rainfall areas. Men must *balance* their minerals, reduce the soil alkalinity with "chemicals," and import special fertilizers from other regions—or from the sea.

As a general rule, you cannot profitably "balance" an off-balance soil of either a low or high rainfall area in a permanent sense. But you can successfully do these things: (a) study your soils and crops, so you may know what is lacking in the soil, and in the tissues of your plant; (b) skillfully fertilize your crops with needed minerals, chelating agents, organic materials, and foliar sprays.

Fertilize the plants and crops; not necessarily the soils. Achieve a mineral balance in the plants if it is too expensive to achieve it in your soils. In this way farmers and gardeners of all climate regions can grow mineral-rich, nutritious, pest-resistant crops without excessive costs.

As described in previous chapters, seaweed and fish have increasing roles as strategic materials in such practical modes of farming and gardening.

WHAT ARE THE ESSENTIAL MINERALS?

To think accurately in this field, it is useful to "think ecological," and also to "think historical."

179

Imagine an ancient plant emerging from seaside ancestors, for example *ponticum rhododendron*. It finds comfortable ways to live in shady places, in the duff of forests, where old bark, needles, mosses, ferns, lichens, rotting limbs, and roots make up the soil; steaming and stewing away, exuding mild organic acids that release copious supplies of iron, zinc, manganese, copper, boron, molybdenum, vanadium, and other minerals for *ponticum* to use. Its neighbors, living helpfully nearby, are devil's clubs, trillium, kalmia, sword ferns, salmon berries, salal, huckleberries, bugs, worms, camp robbers, mycorrhiza, and colonies of clostridia bacteria.

Such a plant, we believe, has built into its tissues and life system *all* of the minerals in its historical environment, and it would be risky to say that only eight, twelve, or twenty are essential. In a *complete* sense the number might be forty or fifty.

Due to its historical environment, *ponticum* is accustomed to supergenerous supplies of iron, manganese, magnesium, boron, and other minerals of rotting vegetation, released effectively by organic acids in a low pH soil, probably about 5.2.

If a landscape nurseryman plants *ponticum rhododendron* in your ornamental border, it may survive, but it will remember the old forest habitat and hunger for ample supplies of iron, manganese, zinc, boron, magnesium, and other minerals; and it will not grow well unless you learn to provide them.

You and the nurseryman have changed and oversimplified the nutrition and environment of this plant; as a consequence, a problem arises: how to feed it so that it is happy in the new place.

What nutrients are essential? Probably all of the forty or more kinds found in forest litter and in the tissues of the plant. The enzyme and hormone systems alone, fully functioning, may require dozens of minerals, although the plant will survive at subpar efficiency on less.

However, *ponticum rhododendron* only illustrates a wild plant with special nutritional needs evolved during eons of time. Let us further consider what minerals are essential when plants are grown for animal feeds and human foods.

THE NUTRITIONAL FACTOR

When plants and crops are to be eaten by animals, poultry, and people, the concepts of essential minerals should be expanded. For example, cobalt and iodine are not considered by U.S. farm research scientists to be essential for growth of plants, but they are recognized to be necessary in the diets of animals and people. Either we get them in our foods or we suffer from malnutrition.

Going a step further, when the U.S. moved most of its vegetable production into long season, hot-weather states (California, Arizona, Texas, and Florida), for rapid growth and succession cropping (two or three crops per year on the same land area), we probably debased these food crops nutritionally. Earlier, when grown in a more leisurely fashion on thousands of farms in many states, the crops could mine the soil for minerals; but now they are grown on simplified diets mainly of nitrogen, phosphate and potash (NPK), with other minerals supplied only to sustain gross yield.

And, as many readers know, you can grow fine-looking high-yield crops which are mediocre as feed for animals and foods for people—demineralized by the mode of farming—and that is what is occurring in the U.S. today. The high yield superefficiency crops are probably inadequate to serve as a foundation for national good health.

Indicating this, the National Nutrition Survey of 1965–1966 showed a widespread occurrence of anemia attributed to low intakes of iron, especially in women, and that finding was corroborated by HEW's Health and Nutrition Examination Survey of 1972, conducted by the Public Health Service. Reacting to this, many nutritionists are suggesting iron supplements, and they have managed to promote "enriched" bread and cereals containing iron and vitamin B supplements.

However, as plant nutritionists, we know that low iron in foods is only a telltale sign of a broad across-the-board mineral deficiency in many crops and products. Zinc, manganese, copper, magnesium, and the whole array of other essential minerals are assimilated by plants, if available, in known ratios to each other; and if iron is low, it is almost certain that

181

zinc, manganese, copper, cobalt, selenium, and others are low, too.

We cannot prove this. The necessary research studies to assess comparative values of food crops grown in different ways have not been conducted. However, in twenty-five years of examining plant tissues, sap, and soils, we have seen evidence of large differences in mineral composition of plants depending on how they were fertilized and grown.

Based on this work, we have an expanded idea of what minerals are essential for growing nutritious food crops, and we suspect that *all of them* have roles of one kind or another in nourishing healthy, disease-resistant plants.

To give an idea of what "all of them" might be, we show the mineral analysis of typical *Ascophyllum nodosum* seaweed, grown in an environment of earth's runoff waters, where it could assimilate all the minerals it wished or needed. Additional analyses of seaweed are shown in the Appendix.

THE PEST CONTROL FACTOR

During Clemson's pioneer work studying seaweed, T. L. Senn and his associates frequently noted that plants fertilized with seaweed had lower insect populations and less pest damage. As Senn said, "The seaweed seemed to change the habits and hunger of insects. For example, flea beetles on tomatoes did not feed voraciously or hop around much but tended to be docile. When touched with a pencil point they often acted lazy and did not hop away, while flea beetles on untreated plants hopped readily, multiplied to large numbers, and damaged the plants."

Glen Graber noted similar effects on insects and plants from seaweed fertilization in four hundred acres of diversified vegetable production in Ohio. The authors have observed the same results on cucumbers, zucchini, and other plants and crops. In fact, worldwide farm research is dotted with scattered reports that seaweed helps plants to resist damage by insects and diseases.

However, until recently there was only a low official interest in such pest control capabilities of seaweed and other

MINERAL ANALYSIS OF SEAWEED
(NORWEGIAN *ASCOPHYLLUM NODOSUM*)

MINERAL ELEMENTS (PERCENTAGES)

Silver	.000004	Sodium	4.180000
Aluminum	.193000	Nickel	.003500
Gold	.000006	Osmium	.000001
Boron	.019400	Phosphorus	.211000
Barium	.001276	Lead	.000014
Calcium	1.904000	Rubidium	.000005
Chlorine	3.680000	Sulphur	1.564200
Cobalt	.001227	Antimony	.000142
Copper	.00635	Silicon	.164200
Fluorine	.032650	Tin	.000006
Iron	.089560	Strontium	.074876
Germanium	.000005	Tellurium	.000001
Mercury	.000190	Titanium	.000012
Iodine	.062400	Thallium	.000293
Potassium	1.280000	Vanadium	.000531
Lantanum	.000019	Tungsten	.000033
Lithium	.000007	Zinc	.003516
Magnesium	.213000	Zirconium	.000001
Manganese	.123500	Selenium	.000043
Molybdenum	.001592	Uranium	.000004

OTHER ELEMENTS PRESENT

Bismuth	Gallium	Thorium
Beryllium	Indium	Radium
Niobium	Iridium	Bromine
Cadmium	Palladium	Cerium
Chromium	Platinum	Rhodium
Cesium		

From Norwegian Institute Of Seaweed Research as reported in *Review Of Seaweed Research*. Research Series No. 76. South Carolina Agricultural Experiment Station. (Clemson, S.C.: Clemson University, 1966).

mineral-rich materials, since it was felt in USDA and elsewhere that *toxic pesticides are cheaper and more effective to use*.

Also, as farm technologists may acknowledge, pest control with toxic chemicals is compatible with large-scale monoculture systems of farming. You simply spray a 300 or 500 acre field of a crop, while *nutritional pest control requires changes in the farming system*. For example, to use seaweed and/or trace mineral controls successfully, the big farmer would probably need to utilize mixed crop farming and crop rotations again, reducing the pest populations and hazards, *then* effectively controlling the pests that remain to damage the crops. This is the experience of Glen Graber: he had to be a good farmer in order to achieve pest control with seaweed.

With the petrochemical age closing, we are entering, as they say, "a new ball game." The costs of toxic pesticides are steadily rising; heavier applications are required to get last year's controls and insect kills; costly chemical pollution of land, water, and foods is occurring; and the contemporary farm chemical industry is moving into a swamp of rising costs and lawsuits.

With another 50 percent rise in costs of parathion, DDT, sevin, 2,4-5T, dieldrin, nematacides, endrin, and other pest poisons, there will surely be a vigorous swing to alternative means of plant protection. Full mineral fertilization of plants and crops is one of the more promising of these alternatives.

Again, we are back to the questions: "What minerals are *essential?*" and "What is complete mineral fertilization of plants?"

And again, we are moving toward the same answer: "Where pest control is concerned, we must consider *all* minerals that are biologically active in life systems on earth, all minerals that *may* have roles in the metabolism of plants." Since plant growth entails the work of literally hundreds of enzymes, each with a mineral or sulphur as a key element in its chemical structure, it is more plausible to postulate that *many* minerals serve in plant-life systems, rather than *just a few*.

Summarizing these thoughts, we find that complete mineral fertilization of plants and crops is desirable because:

184

1. This provides better nutrition of animals and poultry, reducing veterinary costs, and enabling animals and poultry to be better foods for people.

2. It improves nutritional values of human foods, providing a better basis for national health and work efficiency.

3. This is a lower-cost way to obtain pest control in many kinds of crops, especially when used in conjunction with diversified mixed-crop farming.

4. It is a mode of pest control that avoids pollution of land, water, and foods as occurs when toxic pesticides are used.

This raises further questions: "How can it be done at reasonable cost?" and "What kinds of mineral materials should farmers and gardeners use?"

ONLY SMALL QUANTITIES OF MINERALS ARE NEEDED

To zero in on this phase, it is useful to know that most of the minerals involved in plant growth act as catalysts, rather than as cell-building materials, per se; therefore they are not consumed in the process. In *The Diagnosis of Mineral Deficiencies in Plants*, by T. Wallace that principle is stated in the following words (page 13):

> Since catalysts are not used up in the chemical reactions which they promote, we can understand how it comes about that quite small or even minute quantities of the "trace elements," iron, manganese, boron, zinc, and copper, may nevertheless be essential to the plant's health and growth.[1]

Iron may be used as an example. As the most copiously required trace mineral, an iron level of 80 parts per million is nevertheless considered fully adequate in the tissues of many plants. With such an iron content a big hay crop of 5,000 pounds of dry matter per acre would contain less than

[1] T. Wallace, *The Diagnosis of Mineral Deficiencies* (London: Her Majesty's Stationery Office, 1951).

185

8 ounces of iron. If the soil provides half of this, the farmer needs to find ways to provide 4 ounces in his fertilizing program.

In the case of a 120 bushel yield of corn per acre, we are still considering a need for less than one pound of iron. If the soil can provide half of it, the farmer needs to find ways of infusing the other 6 or 8 ounces into his crop; perhaps paying for it by the reduction of his veterinary bills.

Now, iron is the most-needed trace mineral. The "model fertilizer law" and regulations, used in many states, show an optimum relationship between several of the "essential" minerals as follows:

MINERAL	AMOUNT
Iron	.1%
Manganese	.05% = ½ of iron
Zinc	.05% = ½ of iron
Copper	.05% = ½ of iron
Boron	.02% = $^1/_5$ of iron
Molybdenum	.0005% = $^1/_{200}$ of iron
Cobalt	.0005% = $^1/_{200}$ of iron

Enough of all these might be provided for most crops with only 2 pounds per acre of the assorted minerals; and from here we sail off into a miniworld, as far as quantities are concerned. For example, in the case of vanadium, a calculation of the amount actually needed shows that 2 ounces would fertilize about 2,000 acres of crops; something like a "whiff" or a "smell" per acre. The calculation is shown under "Big Numbers" on a page to follow.

In summary, it is evident that only small quantities of minerals would be needed to achieve complete fertilization of plants and crops. However, logistical problems must be solved. In what forms should the minerals be provided? And, how can these fertilizers be applied in efficient ways at low costs?

BIG NUMBERS

(This page is reserved for persons twelve years of age and under whose minds have not been overstuffed with academic knowledge)

Assume 10 billion active cells in a plant: 10×10^{10} cells.

Assume each cell has a particular enzyme that is dependent on vanadium as its key mineral; or some other metal not known to be required by plants. Assume further that each cell has 1 million of these enzyme molecules: 1×10^6 atoms.

Assume 30,000 plants per acre: 3×10^4 plants.

Multiplying these numbers together shows that, under these assumptions, the acre of plants would require: $1 \times 10^{10} \times 1 \times 10^6 \times 3 \times 10^4 = 3 \times 10^{20}$ atoms of vanadium.

However, 51 grams of vanadium (one gram molecular weight) contains 6.02×10^{23} atoms of vanadium, which is a sufficient supply for 2,000 acres. And, this is less than 2 ounces of vanadium.

If there were twice as many cells, and if each cell used 10 times as much vanadium, the 51 grams (2 ounces) would still suffice for 100 acres.

These numbers show the difficulty of determining a plant's needs for trace elements—or, for teensy sniffs of hormones.

However, honorific doctors of agronomy sometimes make synthetic agents, substances and solutions to utilize in research and testing. How difficult it is to do this work! For example, if you wanted to clean down to the last billion atoms of vanadium from a synthetic medium, it would mean the removal of all but:

5×10^{-14} grams $= .00000000000005$ grams

$= 5 \times 10^{-11}$ milligrams

$= 5 \times 10^{-8}$ micrograms

$= 5 \times 10^{-5}$ nanograms

And, believe us, that's not very much material. Where is the genius who will tell us how much of each of seaweed's 50 trace minerals, working together, is useful to plants?—to say nothing about the teensy amounts of hormones.

GOOD MINERAL MATERIALS FOR FARMERS AND GARDENERS TO USE

In many years of full nutrition fertilizer practice, we have employed many different kinds of special materials to supplement nitrogen, phosphorus, and potassium with good assortments of secondary and trace minerals. The following list may be helpful; however, it is not complete and it may omit a good material found in your own neighborhood, such as flue ash or a cannery or food-processing waste.

The rules are: When possible, use composite materials; shotgun feeding rather than rifle shots of individual elements. Use organic wastes, if available, since they often contain sound assortments of minerals in biological ratios. Use sap and plant tissue tests as guides, when possible. And, while informing yourself, do not try to outguess Nature. Provide *everything*, and let the plants pick and choose their foods in cafeteria fashion.

Suggested materials:

• **Good compost or manure.** Start here, because the plant or animal providing these wastes collected a sensible assortment in the first place and built the minerals into its tissues or voided them in the wastes. But remember: (a) compost and manure are only as mineral-rich as the area or feeds from which they came; i.e., high-rainfall areas are lacking in minerals due to leaching for centuries, and this deficiency pervades plants, animals, wastes, and waters, (b) solid manure is notably deficient in phosphorus and potash and some other minerals, and (c) most wastes and composts are unbalanced and incomplete fertilizers. Even seaweed is an unbalanced and incomplete plant food.

• **Lake-bed gypsum and some kinds of marl.** In some areas these are rich in mineral nutrients of their areas, being from the collecting basins and terraces. For example, lake-bed gypsum of the Tonasket area, State of Washington, offers 5% Epsom salts (magnesium sulphate) plus calcium sulphate and a host of other mineral elements.

- **Humates.** Derived from old deposits of vegetation, this coallike substance may have good supplies of supplementary minerals. We like to use it, when available at low cost, as a fertilizer "filler." It's one kind of a mineral "cafeteria."

- **Sulphate of potash-magnesia (Sul-po-mag).** This is a positive supplier of potash and magnesium in a sound ratio, along with sulphur. The official rating is 22 percent potash (K_2O), 18 percent magnesium (MgO) and 27 percent sulphur (S). It's good to add in a fertilizer mixture at about 10 percent of the whole mix (5 percent to 15 percent). Also, this fertilizer may be applied to farm and garden soils at ½ to 1 pound per 100 square feet of area (200 to 400 lbs. per acre). It is basically an old marine salt deposit.

- **Fritted trace elements** (FTE). This product is composed of minerals blended into molten rock (silica), in a fashion simulating Nature (volcanic action), and with the solubility governed by a predetermined hardness of the crystals in the product. It's basically nonsoluble in water, when applied in farming and gardening, or in a fertilizer. A typical FTE product may contain iron, manganese, zinc, copper, boron, and molybdenum in a sensible relationship, or assortment. The FTE may be used at 1 to 2 ounces per 100 square feet of area (30 to 60 pounds per acre), but is best mixed into a fertilizer, so it can be spread evenly.

- **Chelated trace minerals*** One of the best is KE-MIN, using lignin sulphonate as the chelating agent. Plants use both lignin and sulphur, so the product does not adversely affect plants with the chelating substance. It is an "organic" concept. The "multi" KE-MIN contains the same minerals as FTE (Fe, Mn, Zn, Cu, B and Mo), in a reasonable assortment. This product is 100 percent soluble, and may be mixed into foliar sprays and irrigation water, or into a dry fertilizer mix. If KE-MIN is unavailable, ask your farm or garden supply

* A chelated mineral is one that is contained in the chemical structure of an organic molecule, therefore protected from reaction (and loss as a nutrient) with other minerals or compounds.

189

dealer for his recommendation of a chelated mineral product.

- **Blood, seeds, and meat-dressing wastes.** When locally available, these kinds of materials offer good supplies of trace minerals. This is the theory, and the fact: blood of creatures concentrates mineral elements used to support life and therefore can offer such minerals to plants. It is an excellent provider of iron, manganese, zinc, cobalt, copper, and many others; and also contains nearly 16 percent nitrogen. Glands, nerve tissues, and visceral organs have similar though lesser contents. Feathers and hair are concentrators of minerals; in fact, hair analysis of humans is a procedure to check the sufficiency of mineral nutrition. And, hair and feathers are high providers of nitrogen; they are protein-givers. Seeds gather and offer a whole range of foods in a similar fashion. Built to give complete nourishment to young seedlings, they contain whole assortments of minerals; therefore, most kinds of seed meals are good fertilizers from the mineral standpoint.
- **Municipal sludge.** As described in chapter 8, sewage sludge is a major fertilizer resource, for trace minerals as well as nitrogen. However, it is acceptable for growing human foods only when competently processed, and certified to be free of disease pathogens and undesirable heavy metals, such as chronium, mercury, and lead.
- **Fish and fish wastes.** Again, as described in chapter 5, fish materials are superb mineral fertilizers. When derived from marine waters, fish assimilate complete assortments of minerals and offer them to plants in superior forms.
- **And finally . . . seaweed.** Amply discussed in foregoing pages, this is the future major source of complete mineral assortments for land-based food production. However, these characteristics of mineral groups in seaweed should be noted again and sometimes corrected in horticultural practice:

(a) The salt level is high, and can add to problems in saline soils.

190

(*b*) The ratios between Ca, Mg, K, and Na (the mineral cations used in quantity by plants) are a bit haywire. Sodium and potassium are so high that they may immobilize magnesium; and they are, in fact, high in relation to calcium. We think this imbalance of minerals in seaweed may account for reports of decreasing effectiveness when it is applied at over 300 pounds per acre. When properly balanced with other materials, we think it could be used at higher rates.

(*c*) The ratio between manganese and iron is poor, with manganese exceeding iron rather than in the reverse normal role. Also, zinc is low in relation to other elements.

Despite these faults, seaweed is a superior provider of a "whole cafeteria" of minerals for good nutrition of plants. Informed people know how to balance seaweed with other nutrients, thereby getting wholly plus values from it. Also, the minerals in seaweed are organically chelated within the seaweed tissues, and seaweed lends itself to conversion to liquid forms, for use in foliar feeding of plants. It is an excellent trace-mineral fertilizer.

At a future time, *sea solids* may enter the horticultural scene with potential values for providing mineral nutrients, along with seaweed and fish. These are the mineral constituents of seawater when the water has been removed. A brief description of sea solids is given below.

Names and addresses of suppliers of some of the above materials will be found in the Appendix.

THE ART AND SCIENCE OF COMPLETE MINERAL FERTILIZATION

Similar to human beings, when plants desperately need a mineral nutrient (or a group) they may respond to its presence in a dramatic way. For example:

• On an elite golf course in July, in Seattle, Washington, the putting greens had been fed with urea nitrogen repeatedly to groom them for a tournament, thereby inducing an acute need for iron and magnesium with which the grass plants could resume making chlorophyll. As a demonstration, we sprayed *half* of one putting green with a suitable dilute solution of iron, magnesium, and other useful nutrients, and within 30 minutes could photograph the new green color "grown" by the grass in this half we sprayed. The grass was starved for these foods.

• In an orchid greenhouse on Bainbridge Island, State of Washington, we mixed a boron solution for use in making test applications to some selected orchids in a certain greenhouse. We carried this boron solution (quite dilute) through two other greenhouses on the way to the target greenhouse. However, the orchids along this pathway needed boron, and the next day our route was clearly marked by new vigor and green color in these plants that were waiting for a boron supply, which they obtained from the imperceptible fumes from the pail we carried.

Plant and human nutrition are that complex and specific, and the responses are enormous when hungers are adequately attended to. The art and science of all this are to be found in emulating Nature up to a point, then adding cost-saving technologies and materials as they are needed and are compatible with sound ecological principles.

Foliar feeding of plants and crops is such a sound addition to Nature's ways of nourishing her crops. It is a superefficient way to provide small quantities of minerals, while soil feeding may be inefficient and expensive. Phosphate may serve as an example, although it is not a trace mineral; but more is known about phosphate, a stellar nutrient in conventional fertilizer practice.

Agronomic studies show that as a rule, less than 12 percent of phosphate in soil-applied forms is successfully recovered and used by crops. The other 88 percent reacts with other soil and fertilizer elements, gets "tied up," and becomes unavailable to the plants. Iron reacts with other elements in a similar fashion, notably with phosphorus and calcium, and much of it becomes unavailable. Also, an alkaline soil (pH above 7.0) will often tie up and immobilize the whole range of trace minerals as well as iron.

192

Avoiding these hazards and losses, foliar sprays have a cost-saving advantage and a rising place in fertilizer technologies of the future.

The following suggestions illustrate kinds of innovations that may be useful:

- Spray your crops at dawn or in the cool of the evening. The stomata of the leaves are open then and will receive the foods.
- Use a good wetting agent, a surfactant, but remember, it is a "chemical" too. It may feed the plants, itself, or enhance or depress some of your nutrients.
- Add trace minerals to your compost pile. They will first feed the composting bacteria, then your plants, doing double duty.

One year at the Puyallup Agricultural Experiment Station, near Seattle, we painted the insides of flowerpots with minerals, so the roots, upon reaching the edges of their soils, still could be fed. And it worked beautifully. Why are such techniques not being used now? Because the young research genius with whom we were working went away . . . and so did we. It is a bit like answering the classic question: "If you are so damned smart, why aren't you rich?"

It is often a matter of time, place, and where you are on the great curve of history. At present, we are at a certain point: We are in transition from the Petroleum Age to the Energy-Saving Age; too smart for one, not smart enough for the other.

Within thirty years, if the gas wells go dry, we may be dousing pots in minerals, painting leaves with *cytokinins*, and using bees to carry fertilizers to the crops because we are out of gasoline. But mainly we will widen the shores of farm research as T. L. Senn did at Clemson University in 1959, when he commenced studying seaweed.

SMALL-UNIT FARMING IN 1985

(VIGNETTE)

PA: Hey, son, did you fertilize the field?

JR.: Yep, forty pounds of actual nitrogen and I used the coated time-release pellets. Reduces the loss.

PA: Did you put on the compost?

JR.: Yep, 400 pounds per acre to feed the bacteria. And they are chomping on last year's straw and the cover crop, too. That field will be live and sizzlin' in another month.

(time lapse)

PA: Crop's up, son. Are you going to spray?

JR.: Sure. Seaweed and fish at the second joint.

PA: How much?

JR.: Quart per acre. That's enough. *Cytokinins*, you know.

PA: Are you going to add the bio-doozies to take care of the bugs?

JR.: Yep, but the seaweed does most of it.

PA: Son, did you get the government payments yet?

JR.: Yep, $1,500 energy-saving certificate and $500 small-unit bonus. Just $2,000.

PA: Did you pay the fertilizer and spray bill?

JR.: Sure, the payments just covered it.

PA: Did the government man come yet?

JR.: No, he's coming tomorrow and bringing half a dozen fellows. They couldn't get enough fertilizer and they'll look at our fields. See how we did it.

PA: Okay, son. Keep on truckin'.

APPENDIX A.

ANALYSES OF SEAWEED

1. AVERAGE ANALYSIS OF NORWEGIAN SEAWEED MEAL (*ASCOPHYLLUM NODOSUM*)

Moisture content	10.7%
Protein	5.7
Fat	2.6
Fiber	7.0
Ash	15.4

MINERAL ELEMENTS (PERCENTAGES)

Silver	.000004	Sodium	4.180000
Aluminum	.193000	Nickel	.003500
Gold	.000006	Osmium	.000001
Boron	.019400	Phosphorus	.211000
Barium	.001276	Lead	.000014
Calcium	1.904000	Rubidium	.000005
Chlorine	3.680000	Sulphur	1.564200
Cobalt	.001227	Antimony	.000142
Copper	.00635	Silicon	.164200
Fluorine	.032650	Tin	.000006
Iron	.089560	Strontium	.074876
Germanium	.000005	Tellurium	.000001
Mercury	.000190	Titanium	.000012

195

MINERAL ELEMENTS (PERCENTAGES)

Iodine	.062400	Thallium	.000293
Potassium	1.280000	Vanadium	.000531
Lantanum	.000019	Tungsten	.000033
Lithium	.000007	Zinc	.003516
Magnesium	.213000	Zirconium	.000001
Manganese	.123500	Selenium	.000043
Molybdenum	.001592	Uranium	.000004

OTHER ELEMENTS PRESENT

Bismuth	Gallium	Thorium
Beryllium	Indium	Radium
Niobium	Iridium	Bromine
Cadmium	Palladium	Cerium
Chromium	Platinum	Rhodium
Cesium		

From Norwegian Institute Of Seaweed Research, as reported in *Review Of Seaweed Research 1958–1965*. Research Series No. 76. South Carolina Agricultural Research Station. (Clemson, S.C.: Clemson University, 1966).

2. COMPARATIVE ANALYSES OF ICELANDIC SEAWEEDS: DRY SEAWEED MEAL

			LAMINARIA DIGITALIS	ASCOPHYLLUM NODOSUM
Nitrogen	(N)	percent	1.46	1.21
Phosphorus	(P_2O_5)	percent	.57	.38
Potassium	(K)	percent	8.73	3.25
Calcium	(Ca)	percent	1.72	2.63
Magnesium	(Mg)	percent	.87	1.00
Sodium	(Na)	percent	5.56	7.92
Chlorine	(Cl)	percent	9.17	3.98
Iron	(Fe)	parts per million	836	353
Manganese	(Mn)	parts per million	13	21
Zinc	(Zn)	parts per million	15	19

			LAMINARIA DIGITALIS	ASCOPHYLLUM NODOSUM
Copper	(Cu)	parts per million	. 5	3
Cobalt	(Co)	parts per million	.7	2
Iodine	(I)	parts per million	1987	985
pH			6.07	5.31
Ash (salt free)			16.44	17.48

Data from Icelandic Fisheries Laboratories, Reykjavik, Iceland. 1975.

3. ANALYSES OF SOLUBLE SEAWEED POWDER

			SEA BORN * DEHYDRATED SEAWEED POWDER	DEHYDRATED ** POWDER USED IN CLEMSON RESEARCH
Nitrogen	(N)	percent	1.0	.9
Phosphorus	(P)	percent	.07	.09
Potassium	(K)	percent	2.20	2.41
Calcium	(Ca)	percent	.04	.02
Magnesium	(Mg)	percent	.01	.05
Iron	(Fe)	parts per million	147	145
Manganese	(Mn)	parts per million	1	11
Zinc	(Zn)	parts per million	25	23
Copper	(Cu)	parts per million	4	9
Molybdenum	(Mo)	parts per million	—	2.5
pH			9.49	—

* As reported by Icelandic Fisheries Laboratories, Reykjavik, Iceland, 1975.
** As reported by Agricultural Chemical Services, Clemson University, Clemson, South Carolina.

197

4. ANALYSES OF LIQUID SEAWEED EXTRACTS

Note: These data are only indicative of levels of nutrients. Analyses of other samples of liquid seaweed extract may vary from these.

			SAMPLE NO. 1	SAMPLE NO. 2
Nitrogen	(N)	percent	1.27	.069
Phosphorus	(P)	percent	.26	.07
Potassium	(K)	percent	15.24	8.75
Calcium	(Ca)	percent	.30	.90
Magnesium	(Mg)	percent	1.39	1.32
Iron	(Fe)	parts per million	.30	.90
Manganese	(Mn)	parts per million	18	12
Zinc	(Zn)	parts per million	54	25
Copper	(Cu)	parts per million	19	9
Cobalt	(Co)	parts per million	.3	.1
Sodium	(Na)	parts per million	10.18	4.54
pH			4.52	5.29
Ash (salt free)			19.58	29.79

Data from Icelandic Fisheries Laboratories, Reykjavik, Iceland. 1975.

APPENDIX B.

GLOSSARY OF WORDS, TERMS, AND PHRASES

Agar (or Agar-Agar). A colloidal substance derived from seaweed.

Algae. A large group of plants having chlorophyll but no true roots, stems, or leaves, often growing in water. Sea weeds are in this family of plants.

Alginates. The salts of alginic acid, derived from seaweed.

Alginic Acid. A carbohydrate of seaweed. In brown seaweeds, the alginic acid may be 10 percent to 30 percent of the seaweed substance, dry weight. It is used to make gels, thickening, and many food and industrial products.

Auxin. A plant hormone.

Balanced Nutrition of Plants. Modes of fertilizing plants that provide full assortments of nutrients in sound ratios to each other; for example, sufficient potassium and magnesium to "balance" the nitrogen, and sufficient manganese, zinc, iron, and other minerals to "balance" the major nutrients.

Carrageenen. A seaweed substance from which colloids and gels are made for food, medical, and other uses.

Compost. Organic matter more or less converted to humus by action of bacteria.

Cytokinins. A group of plant hormones reputed to stimulate photosynthesis and promote plant growth. *Kinetin* is of this hormone group.

199

Dolomite. A liming material composed of both calcium and magnesium carbonates. A superior kind of lime.

Eelgrass. A grass-type plant that has adapted itself to grow in seashore places, deriving some of its nutrition from seawater.

Energy Crisis. A natural or contrived shortage of energy materials and fuel, resulting in excessive costs, work stoppages, transport problems, and human discomfort.

Enzymes. Specialized protein molecules that serve as regulators, catalysts and digestors in life systems, including those of plants. Using genetic "patterns," enzymes govern cell building and plant growth.

Fish Solubles. A by-product of the fish meal industry, being the soluble component of fish. Since it contains about 5 percent nitrogen (N), 2 percent phosphate (P_2O_5) and 2 percent potash (K_2O), it may be used for fertilizer. However, it is goopy, smelly, and hard to handle.

Foliar Feeding. Fertilizing plants with liquid nutrients sprayed on the leaves.

Fossil Fuels. Coal, petroleum, natural gas, and other nitrogenous substances accumulated on earth from previous animal, marine, or vegetable life.

Fritted Trace Elements (FTE). A trace element fertilizer material made by fusing nutrient elements into a silica or other matrix, using high temperature (about 2,900 degrees F.), and controlling solubility of the end product.

Frond. Branch of a seaweed plant.

Gibberellin. A plant hormone, reputed to assist in governing growth of stems.

Gypsum. Calcium sulphate used to fertilize and improve soils. Especially useful in areas of low rainfall and alkaline soils.

Hardening Plants. Fertilizing to promote maturity and frost resistance, by "ripening" and "hardening" soft, vegetative tissues of the plants. Also, raising carbon versus nitrogen level in the sap to promote fruiting, hardening of the tissues, and frost resistance.

Holdfast. The "anchor" tissues of seaweed plants, attaching them to rocks or debris on the bottom.

Humates. Salts of humic acids.

Humus. The end product of decay of organic matter.

Hydrolyze. To react a substance with water. In the case of fish or seaweed, hydrolyzing breaks down nonsoluble tissues and converts them to water-soluble products.

Kelp. Common name for many kinds of seaweed, and of seaweed foods.

KE-MIN. A chelated trace mineral product based on lignin sulphonate as the chelating agent. It is "organic," water soluble, and compatible with metabolism in plants.

Kinetin. A hormone of the *cytokinin* group.

Liquefied Seaweed. Seaweed tissues that have been treated and pressure cooked (autoclaved) to convert them to liquid form. Since seaweed contains relatively little fibrous tissue, this is rather easily done.

Mannitol. A carbohydrate of seaweed that contains molecules of the alcohol group.

Nematodes. Tiny soil organisms that feed on plant roots causing reduction of growth and crop yields.

NPK Fertilizers. Those containing mainly nitrogen (N), phosphorus (P or P_2O_5) and potassium (K or K_2O), and little or nothing else of the many plant nutrients.

Organiform. A substance formed by reacting and/or combining organic matter with urea and formaldehyde. It is the product of the "organiform process" in treating organic wastes.

Petrochemicals. Chemical products made from petroleum, natural gas, coal, or other fossil substances.

pH. A measure of acidity and alkalinity in any substance, including soils and fertilizers. A pH of 7.0 is neutral, being the level of distilled water. Lower numbers indicate increasing acidity; higher numbers indicate increasing alkalinity.

Plant Nutrients. Minerals, plus nitrogen which is a nonmineral element. They are divided into: *major elements:* nitrogen, phosphorus, and potassium; *secondary elements:* calcium, magnesium, and sulphur; and *trace elements:* such as iron, zinc, manganese copper, boron, and vanadium.

Seaweed, Ascophyllum Nodosum. A large brown seaweed of the *Fucaceae* family. Of the intertidal zone, it has been harvested in Scandinavian countries for many years and has served as the basis for the Norwegian seaweed industry.

Seaweed, Bladder Wrack. A type of brown seaweed with air sacs.

Seaweed, Dulse. An edible plant of the red seaweed family.

Seaweed, Eucheuma. A tropical seaweed used in production of carrageenen, colloids and gels.

Seaweed, Irish Moss. An edible plant of the red seaweed family used in making carrageenin, colloids, and gels.

Seaweed, Laminaria Digitalis. A large brown seaweed of the *Laminariaceae* family. Growing in deeper waters, this variety is used increasingly as a source of alginic acid and alginates.

Seaweed, Laver. An edible plant of the red seaweed family.

Seaweed, Macrocystis. A kind of seaweed growing in fairly deep water and attaining large size. It flourishes along the California coast and is harvested for use in making alginic acid and alginates.

Seaweed, Porphyra. A highly prized edible plant of the red seaweed family.

Seaweed, Sargasso. A floating type of seaweed not anchored to rocks or the ocean bottom by "holdfasts." It flourishes in the Sargasso sea of the Atlantic Ocean.

Seaweed Meal. Seaweed that is dried and pulverized into a meal-like texture. It usually contains less than 12 percent moisture.

Seaweed Powder, Soluble. This is liquefied seaweed, further processed to remove the water. It may be reconstituted, like dry milk solids, to make a liquefied seaweed concentrate for spray-feeding of plants.

Slow-Release Fertilizers. Relatively nonsoluble fertilizers decomposed with the assistance of soil bacteria, so that their foods are slowly and gently available to plants. Examples are: compost, manure, seaweed meal, bone meal, ureaform, and fritted trace elements (FTE).

Soil Bacteria, Azotobacter. A family of soil bacteria that decompose organic matter and capture nitrogen from the air. They thrive at a near-neutral pH range, such as between 6.0 and 7.0.

Soil Bacteria, Clostridia. A family of soil bacteria that decompose organic matter and capture atmospheric nitrogen. They thrive at lower pH levels than azotobacter.

Stripe. Stem of a seaweed plant.

Sulphate of Potash-Magnesia (Sul-po-mag). A fertilizer compound containing about 18 percent magnesia (MgO), 22 percent potash (K_2O) and 27 percent sulphur (S).

Superphosphate (Single). Phosphate fertilizer made by treating ground rock phosphate with sulphuric acid, producing a product with 20 percent P_2O_5 content.

Superphosphate (Triple). Phosphate fertilizer made by treating ground rock phosphate with phosphoric acid, producing a product with 45 percent or 46 percent P_2O_5.

Surfactant. A wetting agent to be used in foliar feeding of plants, improving absorption of liquid nutrients by the leaves.

Ureaform. Ureaformaldehyde, made by reacting urea with formaldehyde. It is a nonsoluble nitrogenous compound containing about 38 percent nitrogen.

APPENDIX C.

DIRECTORY OF SUPPLIERS AND HELPFUL AGENCIES

The following list of suppliers is limited mainly to primary companies selling at the wholesale level. Therefore, to locate retailers in your own city or neighborhood, you may write or call these primary companies.

We have tried to make the list complete and accurate. However, some good companies may have been omitted. If so, write to us and the names and addresses of additional qualified companies will be added in future editions of this book. For this purpose and for supplementary information feel free to write to us at the following address:

Earth Foods Associates
11221 Markwood Drive
Silver Spring, Maryland   20902

SUPPLIERS OF SEAWEED, FISH AND OTHER ENERGY-SAVING FERTILIZERS

Atlantic Laboratories
Boothbay, Maine   04537

Specializes in liquid seaweed, seaweed meal, and ferti-lizer-grade fish meal. A primary manufacturer.

Atlantic & Pacific Research, Inc.
Box 14366
North Palm Beach, Fla.   33408

Distributes SM-3 liquefied seaweed and related products; also, nontoxic pest control materials.

Carpole's, Inc.
Box 32
Garfield, Minnesota   56332

Makes and distributes liquefied fish fertilizers and minced fish-meat products. A primary manufacturer.

Calphos Natural Phosphate
Thompson Sales Company, Inc.
Box 246
Montgomery, Alabama   36101

"Soft" phosphate and related products.

Canton Mills, Inc.
Minnesota City, Minnesota   55959

"Organiform" and a line of natural and organic fertilizers.

Ohio Earth Food, Inc.
13737 Duquette Avenue, N.E.
Hartville, Ohio   44632

Seaweed meal, rock phosphate and other fertilizers and pest controls.

Eco-Systems
2011 N.E. 58th Avenue
Des Moines, Iowa   50313

Liquid fertilizers and related products.

The Organic Farm Center
193 Marinwood Avenue
San Rafael, California   94903

Humates, kelp products and other materials.

Maxicrop, U.S.A.
Box 964
Arlington Heights, Illinois   60006

Seaweed meal, soluble seaweed, and other products.

Garden Way Associates
299 Westport Avenue
Norwalk, Connecticut   06851

>   Seeds, tools, and materials for energy-saving farming and gardening.

Jarl Enterprises
Box 151
Millbrae, California   94030

>   Liquid seaweed and related products.

Organic Sea Products Corporation
1550 Rollins Road
Burlingame, California   94010

>   Liquid seaweed and other seaweed and fish products for foods as well as horticulture.

Earth & Sea Products, Inc.
Box 1305
Watsonville, California   95076

>   Composts, "live" earth, seaweed and related products.

Steve Carlsen & Associates
4767 Candleberry Street
Seal Beach, California   90740

>   Compost, seaweed, and related products.

Naremco, Inc.
Box 1572 S.S.S.
Springfield, Missouri   65805

>   Seaweed meal and products for veterinary and horticultural use.

Sea Born, Inc.
2000 Rockland Road
Charles City, Iowa   50616

>   Liquefied seaweed, seaweed meal, fish, and blends of seaweed and fish. A primary manufacturer.

Zook & Rank
Gap, Pennsylvania   17527

Seaweed, composts, and a full line of natural and energy-saving fertilizer materials. A mixer manufacturer.

Frit Industries, Inc.
Box 1234
Ozark, Alabama   36360

Fritted trace elements (FTE) and related products. A primary manufacturer.

Georgia-Pacific Company
Bellingham, Washington   98225

KE-MIN chelated minerals. A primary manufacturer.

Kaser Construction Company
Box 3569
Des Moines, Iowa

Gypsum for farm and garden use.

Sudbury Laboratories
572 Dutton Road
Sudbury, Massachusetts   01776

Liquid seaweed and fish fertilizers.

Sea Weed Products
Route 6, Box 6164
Bainbridge Island, Washington   98110

Seaweed meal, liquid seaweed, and related products.

Invivo Corp.
Box 2357
Iowa City, Iowa   52240

A complete line of seaweed materials and products.

Zapata-Haynie Corp.
5010 York Road
Baltimore, Maryland   21212

Fish solubles.

Leisure Group
Union Bank Square
Los Angeles, California   90071

Composts, potting mixes and seaweed materials.

Farm Guard Products
6101 Candelaria Street, N.E.
Albuquerque, New Mexico   87110
> Humates and related products.

Ferma-GRO corporation
Storm Lake, Iowa   50588
> Liquid seaweed-based fertilizers.

Pan Organic Humates
Box 9411
Raytown, Missouri   64133
> Seaweeds, humates and fish fertilizers. Also seaweed and fish blends.

Wonder Life Corporation
4931 Douglas Avenue
Des Moines, Iowa   50310
> Fertilizers and soil treatment materials.

General Compost Corporation
322 South 16th Street
Philadelphia, Pennsylvania   19102
> Pfeiffer compost materials and products.

Medina Agriculture Products, Inc.
Box 309
Hondo, Texas   78861
> Humates and related products.

International Mineral & Chemical Corp.
IMC Plaza
Libertyville, Illinois   60048
> Sulphate of potash-magnesia (Sul-po-mag).

Orgonics, Inc.
Box 543
Slatersville, Rhode Island   02876
> "Organiform" products.

H. J. Baker Bros.
360 Lexington Avenue
New York City   10017

"Organiform," bone meal, fish solubles, and other primary fertilizers.

SERVICE AGENCIES AND ORGANIZATIONS

Unfortunately, the U.S. Department of Agriculture Extension Service is relatively uninformed concerning uses of seaweed, fish, foliar sprays and similar materials for farming and gardening; therefore, it may provide little or no service when called. However, some of the County and Area Extension Agents may be helpful. They may be found in local phone directories under State or County listings, usually under "Agricultural Extension Service" or "Extension Service."

We suggest you try them, particularly for help in composting and handling organic wastes.

In addition, the following agencies and organizations may be useful:

Office of Sea Grant
National Oceanic & Atmospheric Administration
Department of Commerce
Page Building
3300 Whithaven Street
Washington, D.C.   20235

Contact them for information on seaweeds, fish, and marine subjects.

Horticulture Department
Clemson University
Clemson, South Carolina   29631

This is T. L. Senn's research institution. They will respond to questions on uses of seaweed in farming and gardening.

Agronomy Department
University of Maryland
College Park, Maryland   20742

This is John Hall's research base for work with seaweed. They may be helpful as their work becomes further advanced.

209

ACRES, U.S.A.
10227 East 61st Street
Raytown, Missouri   64133

> This is America's "Voice for Eco-Agriculture," an outstanding rural monthly news magazine. ACRES will respond to questions and give guidance on uses of seaweed, fish, minerals and other energy-saving fertilizers.

Rural Advancement Fund
2128 Commonwealth Avenue
Charlotte, North Carolina   28205

> This nonprofit organization conducts practical research and training in energy-saving horticulture, and is acquainted with seaweed, fish, and foliar feeding of plants and crops. They will respond to questions.

Gardens for All
163 Church Street
Burlington, Vt.   05401

> This nonprofit organization will respond to questions on uses of seaweed, fish, and related materials in gardening and horticulture.

Garden Way Laboratories
Charlotte, Vermont   05445

> This laboratory has a working knowledge of seaweed and marine materials and will respond to correspondence. It can provide practical guidance and testing services.

Earth Foods Associates
11221 Markwood Drive
Silver Spring, Maryland   20902

> This is the business and work base of the authors. We shall respond to correspondence and may provide or arrange for services in fields of energy-saving horticulture.

APPENDIX D.

WEIGHTS AND MEASURES

America is moving from the United States-British System of weights and measures to the metric system. The following factors will assist in using either system, and converting from one to the other.

WEIGHTS

	NUMBER OF GRAMS	EQUALS THIS QUANTITY OF WATER	AVOIR- DUPOIS WEIGHT
Metric ton	1,000,000 =	1 cubic meter	= 2204.6 lbs.
Quintal	100,000	1 hectoliter	220.46 lbs.
Myriagram	10,000	10 liters	22.046 lbs.
Kilogram or kilo	1,000	1 liter	2.2046 lbs.
Hectogram	100	1 decliter	3.5274 oz.
Decagram	10	10 cubic centimeters	0.3527 oz.
Gram	1	1 cubic centimeter	15.432 grs.
Decigram	.1	.1 cubic centimeter	1.5432 grs.
Centigram	.01	10 cubic millimeters	0.1543 grs.
Milligram	.001	1 cubic millimeter	0.0154 grs.

AREAS, LENGTHS, AND VOLUMES

1 hectare = 10,000 sq. meters = 2.47 acres
1 are   =   100 sq. meters = 119.6 sq. yds.
1 centare =   1 sq. meter = 1550 sq. ins.

2.54 centimeters = 1 inch
30.48 centimeters = 1 foot
1 meter   = 3.2 feet
1 meter   = 39.37 inches
1 kilometer   = 3280 feet
1 kilometer   = .621 mile

1 kiloliter   = 264.2 gallons
1 decaliter   = 2.64 gallons
1 liter   = 1.056 quarts

1 acre   = 43,560 sq. feet
1 acre   = 208 x 208 feet (approx.)
1 cubic foot   = 1728 cubic inches
1 cubic foot   = 7½ gallons
1 cubic yard   = 27 cubic feet
1 pint   = 16 fluid ounces
1 quart   = 32 fluid ounces
1 gallon   = 128 fluid ounces
2 tbsps.   = 1 fluid ounce
2 tbsps. to gal. = About 1 part to 100 parts

FERTILIZER MEASURES (APPROXIMATE)

POUNDS PER ACRE	EQUAL AMOUNT PER 100 SQUARE FEET
100	3 ounces
300	11 ounces
500	1¼ pounds
800	2 pounds
1000	2½ pounds
2000	4¾ pounds

TO MEASURE DRY FERTILIZER *

1 rounded tablespoon	= 1 ounce (approx.)
½ cup	= 3 ounces (approx.)
1 cup	= ½ pound
1 pint	= 1¼ pounds
1 gallon	= 10 pounds

* Compost, dried manure, or peat moss is about 50 percent lighter than this, so more would be used.

RATES TO APPLY FERTILIZERS

Note: These are general guides. Read the product labels and follow their directions.

MATERIAL	PER ACRE	PER 100 SQUARE FEET
Seaweed meal	300 to 500 lbs.	1 lb.
Seaweed powder (reconstituted)	12 oz.	spray until wet
Seaweed liquid concentrate	1 qt. to 1 gal.	spray until wet
Seaweed/fish liquid concentrate	1 qt. to 1 gal.	spray until wet
Fish, liquid	1 gal.	spray until wet
Compost, medium dry	2,000 lbs.	5 lbs.
Manure medium dry	2,000 lbs.	5 lbs.
Manure fairly wet	10,000 lbs.	25 lbs.
Lime (agricultural) * humid areas	2,000 lbs.	5 lbs.
Gypsum (agricultural) * dry areas	1,000 lbs.	2½ lbs.

* Farmers should get soil analyses to determine lime and gypsum requirements.

213

GARDENERS' GUIDES (APPROXIMATE)

A medium-sized handful of medium-weight commercial fertilizer equals about	2½ ozs.
6 medium-sized handfuls of medium-weight fertilizer equal about	1 lb.
A 1-lb. coffee can of medium-weight fertilizer equals about	2½ lbs.
1¼ cups lime or dolomite equal	1 lb.
1 quart dried manure equals	1 lb.
1¼ cups rock phosphate equal	1 lb.
3 cups seaweed meal equal	1 lb.
1½ cups bone meal equal	1 lb.
2 cups organic fertilizer equal	1 lb.
1¼ cups semiorganic fertilizer equal	1 lb.

INDEX

215